Handbook of Manufacturing System

Handbook of
Manufacturing System

Edited by **Jeff Hansen**

LANRYE
INTERNATIONAL

New Jersey

Published by Clanrye International,
55 Van Reypen Street,
Jersey City, NJ 07306, USA
www.clanryeinternational.com

Handbook of Manufacturing System
Edited by Jeff Hansen

International Standard Book Number: 978-1-63240-277-6 (Hardback)

Printed in the United States of America.

Contents

Preface   VII

Chapter 1   **Environmental Burden Analyzer for Machine Tool Operations and Its Application**   **1**
Hirohisa Narita

Chapter 2   **Modelling and Implementation of Supervisory Control Systems Using State Machines with Outputs**   **15**
Moacyr Carlos Possan Junior and André Bittencourt Leal

Chapter 3   **Digital Manufacturing Supporting Autonomy and Collaboration of Manufacturing Systems**   **37**
Hasse Nylund and Paul H. Andersson

Chapter 4   **Reliability Evaluation of Manufacturing Systems: Methods and Applications**   **61**
Alberto Regattieri

Chapter 5   **Integrated Process Planning and Scheduling and Multimode Resource Constrained Project Scheduling: Ship Block Assembly Application**   **77**
Raad Yahya Qassim

Chapter 6   **Migrating from Manual to Automated Assembly of a Product Family: Procedural Guidelines and a Case Study**   **85**
Michael A. Saliba and Anthony Caruana

Chapter 7   **Stochastic Capacitated Cellular Manufacturing System Design with Hybrid Similarity Coefficient**   **115**
Gökhan Eğilmez, Gürsel A. Süer and Orhan Özgüner

Chapter 8   **Stochastic Multi-Stage Manufacturing Supply Chain Design Considering Layered Mini-Cellular System Concept**   **139**
Jing Huang and Gürsel A. Süer

Chapter 9   **Facility Layout**   **159**
             Xiaohong Suo

Chapter 10   **Real-Time Petri Net Based Control System**
             Design for Distributed Autonomous Robotic
             Manufacturing Systems   **187**
             Gen'ichi Yasuda

             Permissions

             List of Contributors

Preface

The aim of this book is to provide updated information regarding manufacturing systems and discuss the current advances, devices, methods and novel ideas in manufacturing systems. Manufacturing system consists of equipment, products, people, knowledge, control and support functions so as to create a framework which caters to market needs and demands in today's competitive world. It collects various researches conducted in fundamental and applied industrial fields. The book will be useful to those involved in manufacturing engineering, systems and management and in manufacturing research.

Various studies have approached the subject by analyzing it with a single perspective, but the present book provides diverse methodologies and techniques to address this field. This book contains theories and applications needed for understanding the subject from different perspectives. The aim is to keep the readers informed about the progresses in the field; therefore, the contributions were carefully examined to compile novel researches by specialists from across the globe.

Indeed, the job of the editor is the most crucial and challenging in compiling all chapters into a single book. In the end, I would extend my sincere thanks to the chapter authors for their profound work. I am also thankful for the support provided by my family and colleagues during the compilation of this book.

Editor

Environmental Burden Analyzer for Machine Tool Operations and Its Application

Hirohisa Narita
School of Health Sciences, Fujita Health University
Japan

1. Introduction

The manufacturing technologies have been evolving according to development of evaluation methods for quality, productivity and cost. In spite of the need for eco-manufacturing, the relationship between manufacturing technologies and environmental burdens has not been yet revealed. Machining is a major manufacturing activity, hence an analyzer developed evables to estimate the environmental burden due to machine tool operations. The machine tool operations has a big potential influence regarding envirnmental burden, and then the envirnmental burden analyzer is developed based on LCA (life cycle assessment) (SETAC, 1999).

Some environmental burden analyses for machine tool operations have the problem which isn't suited to evaluating cutting conditions in detail (Shimoda, 2000, Touma et al., 2003). For example, the conventional methods can evaluate only the difference among dry, wet and MQL (minimal quantities of lubricant) machining operations, but not the difference among depth of cuts, feed rate, spindle speed and tool path pattern. Furthermore, if removal volume and material type are same, the environmental burden becomes same. That is to say accurate environmental burden can't be provided for deciding the cutting conditions.

One research proposed manufacturing planning with consideration of multi-endpoint environmental effects, but any concrete evaluation ways of machining operation haven't been presented (Hara et al., 2005, Sheng et al., 1998). The other researches discussed environmental burden based on energy consumption(Diaz et al., 2010), but did not cmprehensicely evaluate the related environmental burden. I should thus be able to develop the envirnmental burden analyzer for machine tool operations to realize sustainable manufacturing (OECD, 2009). I also proposed a decision method of cutting conditions to achieve minimum environmental burden with using the analyzer developed. The analyzer will enable to accelerate the development of environmental technologies and eco-industries. A calculation algorithm of environmental burden, a system overview and some application example are described in this paper.

2. Life Cycle Assessment

Generally, LCA is a very useful methodology for estimating the environmental burden of a product or service associated with all stages: row-material production, manufacture, transportation, use, repair and maintenance, and disposal or recycling. LCA has four processes: goal and scope definition, inventory analysis, impact assessment, and interpretation, and then realizes holistic assessment of environmental aspect and helps a more informed decision regarding product design modifications and business strategies.

This paper proposes an application technique of LCA to machining processes. For this purpose, machining process and machine tool models in computer environment is constructed and an environmental burden analyzer is developed. Cutting conditions achieving low environmental burden are also discussed by using the analyzer developed.

3. System overview

Figure 1 explains an overview of the environmental burden analyzer for machine tool operations. A workpiece, cutting tool models and an NC program are entered to the analyzer, all activities including a machine tool operation and a machining process are made an estimation. At that time, electric consumption of a machine tool peripheral devices and motors, cutting tool's wear, coolant quantity, lubricant quantity, metal chip quantity and other factors are calculated. Here, the other factors correspond to electric consumptions of air conditioning, light, AGV's transportation, products washing and etc. Using these calculated factors, emission intensity data and resource data, the environmental burden is obtained, when a part is machined. The emission intensity is the rate of an emission matter for an impact category. For example, quantity of carbon dioxide emitted per joule of energy produced for global warming. The emission intensity data is prepared according to an impact category such as global warming, acidification, toxicity to ecosystem, toxicity to human, eutrophication and nuclear radiation. The resource data also is a machine tool specification data, cutting tool parameters and physical parameters of the cutting force for the estimation of machining process. The cutting tool parameter corresponds to tool's diameter, helical angle, rake angle and number of tooth.

This analyzer can calculate the environmental burden in various cutting conditions, because a machining process is evaluated properly. This is a novel aspect of this research as compared with the conventional approach.

Generally, an impact category must be de determined and relevant emission factors of the impact category are selected for LCA. Global warming is determined as an impact category and carbon dioxide (CO_2), methane (CH_4) and dinitrogen monoxide (N_2O) are selected as the it's relevant emission factors. Influences of halocarbon, and sulfur hexafluoride (SF6) on global warming are well known. However, these relevant emission factors are ignored, because their emission intensities have not been found according to my survey. All emissions are converted to equivalent CO_2 emission by multiplying them by characterization factors, and then total equivalent CO_2 emission is calculated as the environmental burden. The global warming potential (GWP) of 100-year impact (IPCC, 2007) is used as the characterization factors, as shown in Table 1.

	CO_2	CH_4	N_2O
Global warming potential (GWP)	1	25	298

Table 1. Characterization factors of global warming (IPCC, 2007)

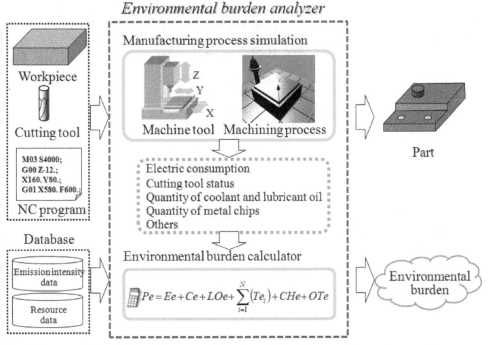

Fig. 1. An overview of the environmental burden analyzer for machine tool operations

4. A calculation algorithm of environmental burden

4.1 Total environmental burden due to a machine tool

Eq.(1) describes total environmental burden due to a machine tool operation, which is obtained from the electric consumption of the machine tool, the coolant, the lubricant oil, the cutting tool's wear, the metal chip and the other factors.

$$Pe = Ee + Ce + LOe + \sum_{i=1}^{N}\left(Te_i\right) + CHe + OTe \tag{1}$$

Pe: Environmental burden of machining operation [kg-CO_2]
Ee: Environmental burden of machine tool component [kg-CO_2]
Ce: Environmental burden of coolant [kg-CO_2]
LOe: Environmental burden of lubricant oil [kg-CO_2]
Te: Environmental burden of cutting tool [kg-CO_2]
CHe: Environmental burden of metal chip [kg-CO_2]
OTe: Environmental burden of other factors [kg-CO_2]
N: Number of cutting tool used in an NC program

OTe isn't introduced in this paper. Calculation algorithms of these factors (Ee, Ce, LOe, Te and CHe) are introduced in detail as follows.

4.2 Electric consumption of machine tool (*Ee*)

Ee means the environmental burden due to the electric consumption in an NC program. Figure 2 shows the electric consumption model of a machine tool, and the environmental burden due to the electric consumption of machine tool is described as follows.

$$Ee = k \; (SME+ SPE +SCE+CME +CPE+TCE1+TCE2+ATCE+MGE+OAE+COE+CUE+SBE) \quad (2)$$

k: CO_2 emission intensity of electricity [kg-CO_2/kWh]
SME: Electricity consumption of servo motors [kWh]
SPE: Electricity consumption of a spindle motor [kWh]
NCE: Electricity consumption of an NC controller [kWh]
SCE: Electricity consumption of a cooling system of spindle [kWh]
CME: Electricity consumption of a compressor [kWh]
CPE: Electricity consumption of a coolant pump [kWh]
TCE1: Electricity consumption of a lift up chip conveyor [kWh]
TCE2: Electricity consumption of a chip conveyor in machine tool [kWh]
ATCE: Electricity consumption of an auto tool changer (ATC) [kWh]
MGE: Electricity consumption of a tool magazine motor [kWh]
OAE: Electricity consumption of an oil air compressor [kWh]
COE: Electricity consumption of an oil mist compressor [kWh]
CUE: Electricity consumption of a chip air blow compressor [kWh]
SBE: Stand-by energy of a machine tool [kWh]

In Eq. (2), the electric consumption of peripheral devices such as an NC controller, a cooling system of spindle, a compressor, a coolant pump, a lift up chip conveyor, a chip conveyor in machine tool, an ATC, a tool magazine motor and stand-by energy are calculated by their running times. But *CPE*, *TCE1*, *TCE2*, *COE* and *CHE* may not be used according to a machine tool operation with or without coolant usage. We must survey through the

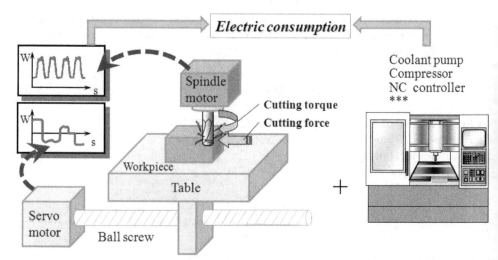

Fig. 2. Electric consumption model for machine tool

machine tool specification previously whether or not to consider these electric consumptions. The electric consumption of the servo motors and spindle motor is also varied dynamically according to the machining process, hence new analysis model must be constructed.

The load torques of servo motors are calculated as follows.

$$T_{L\,servo} = T_U + T_M \tag{3}$$

$T_{L\,servo}$: load torque of servo motor [N·m]
T_U: axis friction torque [N·m]
T_M: application torque of ball screw [N·m]

Here, T_U is a torque due to rubber sealing and can't be obtained theoretically, thus its value is decided by an experiment. T_M is calculated as follows.

$$T_M = \frac{(\mu \cdot M \mp f) \cdot l \cdot \cos\theta \pm (M - f) \cdot l \cdot \sin\theta}{2\pi \cdot \eta} \tag{4}$$

μ: Friction coefficient of slide way
η: Transmissibility of ball screw system
l: Ball screw lead [m]
M: Moving part weight (table and workpiece) [N]
f: Cutting force in an axis [N]
θ: Gradient angle from horizontal plane [rad]

This equation is reconstructed by a monitoring method for cutting force (Fujimura and Yasui, 1994) with normal and reverse rotations of the servo motor. θ is 0 in the X- and Y-axes, and $\pi/2$ in the Z-axis. The cutting force in an axis, f, and the load torque of the spindle motor, $T_{L\,spindle}$, are calculated from the cutting force model. Virtual machining simulator I devloped (Narita et al., 2006) is applied for the purpose of estimating the aforementioned cutting force and cutting torque.

The calculated motor torque is converted to electric consumption as follows.

$$P = \frac{2\pi}{60} \times n \times T_L \times t \tag{5}$$

P: Electric consumption [Wh]
T_L: Load torque ($T_{L\,servo}$ or $T_{L\,spindle}$) [N·m]
n: Motor rotation speed [rpm]
t: Time [hr]

The electric consumption of the compressors corresponding to CPE, TCE1, TCE2, COE and CHE, is also varied according to a discharge pressure and the air dryer, hence the following equation is adapted for the estimation.

$$Ec = CEn + CF + ADFr + ADFa \tag{6}$$

Ec: Electric consumption related to compressor [kW]

CEn: Electric motor power of main motor for compressor [kW]
CF: Rated electric motor power of fan motor for compressor [kW]
ADFr: Rated power of cooling machine of air-dryer [kW]
ADFa: Rated power of cooling fan of air-dryer [kW]

4.3 Coolant (*Ce*)

Ce means the environmental burden due to the coolant in an NC program. There are two types cutting fluid, hence two equations are proposed for *Ce*. Regarding water-miscible cutting fluid, coolant is generally used to enhance machining performance and circulated in a machine tool by coolant pump until coolant is made replacement. During this period, some coolants are eliminated by adhesion to metal chips, hence coolant is supplied for this compensation. The reduction of the dilution fluid (water) due to vapor has to be also considered to estimate total coolant. Here, the following equation is adopted to calculate the environmental burden due to the coolant.

$$Ce = \frac{CUT}{CL} \times \{(CPe + CDe) \times (CC + AC) + WAe \times (WAQ + AWAQ)\} \tag{7}$$

CUT: coolant usage time in an NC program [s]
CL: mean interval of coolant update [s]
CPe: environmental burden of cutting fluid production [kg-CO_2/L]
CDe: environmental burden of cutting fluid disposal [kg-CO_2/L]
CC: initial coolant quantity [L]
AC: additional supplement quantity of coolant [L]
WAe: environmental burden of water distribution [kg-CO_2/L]
WAQ: initial quantity of water [L]
AWAQ: additional supplement quantity of water [L]

4.4 Lubricant oil (*LOe*)

LOe means the environmental burden due to the lubricant oil in an NC program. The lubricant oil is used for two main types. One is for a spindle, another is for a slide way. Here, Minute amounts of oil are supplied by a pump to the spindle and the slide way within a specific interval. Grease as a lubricant is not introduced here, but the same equations can be applied to calculate the environmental burden due to grease. The environmental burden due to the lubricant oil is calculated as follows.

$$LOe = \frac{SRT}{SI} \times SV \times (SPe + SDe) + \frac{LUT}{LI} \times LV \times (LPe + LDe) \tag{8}$$

SRT: Spindle runtime in an NC program [s]
SV: Discharge rate of spindle lubricant oil [L]
SI: Mean interval between discharges [s]
SPe: Environmental burden of spindle lubricant oil production [kg-CO_2/L]
SDe: Environmental burden of spindle lubricant oil disposal [kg-CO_2/L]
LUT: Slide way runtime in an NC program [s]
LI: Mean interval between supplies [s]

LV: Lubricant oil quantity supplied to slide way [L]
LPe: Environmental burden of slide way lubricant oil production [kg-CO_2/L]
LDe: Environmental burden of slide way lubricant oil disposal [kg-CO_2/L]

4.5 Cutting tool (*Te*)

Te means the environmental burden due to the cutting tools in an NC program. All cutting tools are managed by tool life, hence a tool life is compared with machining time to calculate the environmental burden in a machining. Cutting tools, especially for solid end mills, are often renewed by regrinding. The environmental burden due to the cutting tool is calculated as follows by considering the aforementioned processes.

$$Te = \frac{MT}{\sum_{j=1}^{RN+1} TL_j} \times \left((TPe + TDe) \times TW + RN \times RGe \right) \tag{9}$$

MT: Machining time [s]
TL: Tool life [s]
TPe: Environmental burden of cutting tool production [kg-CO_2/kg]
TDe: Environmental burden of cutting tool disposal [kg-CO_2/kg]
TW: Tool weight [kg]
RGN: Total number of regrinding processes
RGe: Environmental burden of regrinding [kg-CO_2]

4.6 Metal Chips (*CHe*)

CHe means the environmental burden due to the metal chips in an NC program.

Metal chip recycling, in which a chip compactor, a chip crusher, a centrifugal separator and an arc furnace are used, generates an environmental burden. In this research, the environmental burden is calculated with considering chip weight as follows.

$$CHe = (WPV - PV) \times MD \times WDe \tag{10}$$

WPV: Workpiece volume [cm³]
PV: Product volume [cm³]
MD: Material density of workpiece [kg/cm³]
WDe: Environmental burden of metal chip processing [kg-CO_2/kg]

4.7 Output example of the developed analyzer

Figure 3 shows an output example of the developed analyzer. The left part shows the machining process instructed by an NC program and estimate cutting force, cutting torque, electric consumption, quantity of cutting oil, quantity of lubricant oil, usage time of cutting tool and metal chip volume. The right part also shows the environmental burden calculated by the aforementioned algorithm. This analyzer can evaluate various environmental burdens by inputting the emission intensities related to the impact categories.

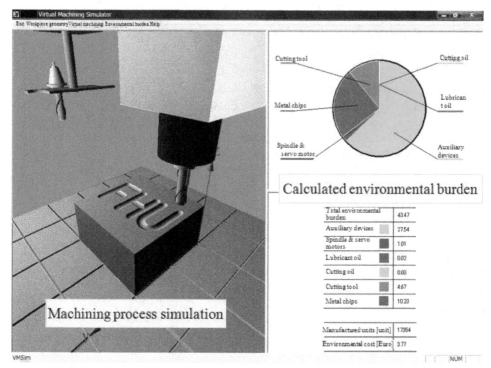

Fig. 3. An output example of the developed analyzer

5. Case study

5.1 Comparison of cutting conditions

For this case study, a machine tool is a vertical machining center (MB-46VA, OKUMA Corp.), a cutting tool is carbide-square end mill with 6mm-diamater and 2-flulte, and a workpiece is a medium carbon steel (S50C) and a compressor is a screw compressors (SCD-110JC, Anest Iwata Corp.).

The parameters to calculate the electric consumption of the servo motor of the machine tool and the electric powers of the peripheral devices of the machine tool are summarized in tables 3 and 4, respectively. These values have been measured and obtained from an instruction manual of the machine tool. The friction torque of servo motors also has been determined by experiment in advance. Table 4 shows the other parameters regarding a machine tool operation.

Table 5 shows the CO_2 emission intensities required to calculate the environmental burden of machining operations. These values were cited from some reports, such as environmental reports, technical reports, homepages and industrial tables (Tokyo Electric Power Company, (2005, Bureau of Waterworks Tokyo Metropolitan Government, 2003, Nansai et al., 2002, Osaka prefecture, 2003, 2005, Mizukami, 2002).

Table part weight [kg]	X: 903, Y:230, Z: 512
Friction coefficient of slide way	All axes: 0.01 (linear guide)
Ball screw lead [mm]	X, Y: 20, Z: 16
Transmissibility of ball screw system	0.95

Table 2. Parameters to calculate the electric consumption of the servo motor

NC controller [kW]	0.16
Cooling system of spindle [kW]	0.45
Compressor [kW]	1
Coolant pump [kW]	0.25
Lift up chip conveyor[kW]	0.1
Chip conveyor in machine tool [kW]	0.6
ATC [Wh]	0.08
Tool magazine [Wh] (1 round)	0.087
Vampire (stand-by) energy [kW]	0.64

Table 3. Electric consumptions and powers of machine tool

Initial cutting fluid quantity [L]	8.75
Additional supplement of cutting fluid [L]	4.3
Initial dilution fluid quantity [L]	175
Additional supplement of dilution fluid [L]	82.25
Mean interval between replacements of coolant in pump [Month]	5
Discharge rate of spindle lubricant oil [mL]	0.03
Mean interval between discharges for spindle lubrication [o]	180
Lubricant oil supplied to slide way[mL]	228
Mean interval between supplies [hour]	2000
Tool life [s]	5400
Total number of regrinding processes	2
Material density of cutting tool [g/cm³]	11.9
Material density of workpiece [g/cm³]	7.1
Coolant tank capacity of machine tool [L]	175

Table 4. Other parameters for machine tool operation

Electricity [kg-CO_2/kWh]	0.381
Cutting fluid production [kg-CO_2/L]	0.978
Cutting fluid disposal [kg-CO_2/L]	0.0029
Dilution liquid (water) [kg-CO_2/L]	0.189
Spindle and slide way lubricant oil production [kg-CO_2/L]	0.469
Spindle and slide way lubricant oil disposal [kg-CO_2/L]	0.0029
Cutting tool production [kg-CO_2/kg]	33.7
Cutting tool disposal [kg-CO_2/kg]	0.0135
Regrinding [kg-CO_2/number]	0.0184
Metal chip processing [kg-CO_2/kg]	0.322

Table 5. Equivalent CO_2 emission intensities

Figure 4 shows a part shape and a tool path pattern used for an example and table 6 shows the cutting conditions of NC program. Three cutting condition: Program 1, Program 1 with coolant (water miscible type) and Program 2 are evaluated. Here, a feed rate of immersion to workpiece is 100 mm/min and the tool life is assumed to be increased to 1.5 times of the original one due to the coolant effect.

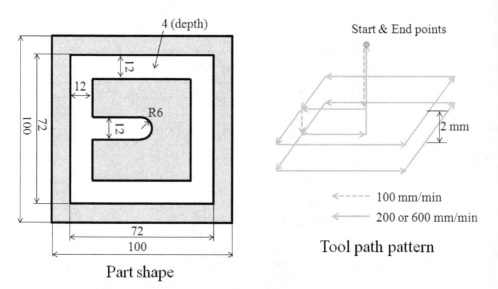

Fig. 4. Part shape and tool path pattern

	Program 1	Program 2
Spindle speed [rpm]	2500	7500
Feed rate [mm/min]	200	600

Table 6. Cutting conditions of NC programs

Figure 5 shows calculated results of three cutting conditions. Program 2 is best in three cutting conditions, because the machining time is very short. The environmental burden due to the cutting tool is reduced by the coolant effect, but the one due to electric consumption of peripheral devices are increased by the usage of coolant pump. As a matter of course, the one of coolant is increased but small. It is found that main reason of the increase of environmental burden due to the coolant usage is the one due to the peripheral devices as shown in this figure. As shown in this case study, the developed analyzer can evaluate various cutting conditions in details.

5.2 Determination method to realize low environmental burden

The environmental burden due to the peripheral devices must be reduced in order to reduce total environmental burden as shown in Fig.5. The one due to the peripheral devices is proportional to time, hence high speed milling might be effect.

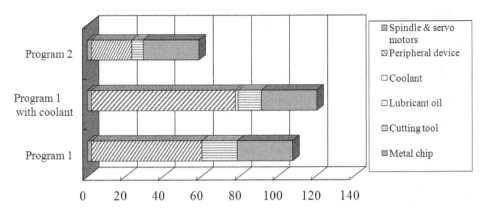

Fig. 5. Comparison of various cutting conditions

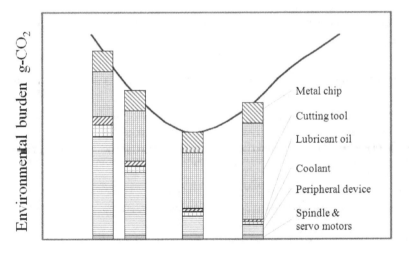

Fig. 6. Tendency due to high speed milling

Here, the relationship between the environmental burden and the cutting speed (spindle speed) is discussed when a feed per tooth, a radial depth and an axial depth of cuts are constant. Figure 6 shows a tendency due to the high speed milling. A tool wear becomes extremely large [Jiang, 2011] and a tool life will shorten in high speed millings, hence the environmental burden due to the cutting tool will increase. But the one due to the electric consumption, coolant and lubricant oil is proportional to time. That is to say there is a trade-off relation between the one due to the cutting tool and the one due to the electric consumption, coolant, lubricant oil. However the one due to the spindle and servo motors is very small (it is constant when the feed per tooth is constant), and the one due to the metal chip is constant, and then these environmental burdens are ignored for discussing. As shown in fig.6, there is a cutting condition to realize the minimum environmental burden. Hence, an optimum cutting speed (spindle speed) can be obtained automatically by calculating an approximate equation with using least-square method and exploring the cutting conditions achieving the minimum environmental burden with using iterative calculation. An optimum cutting speed (spindle speed) is attempted to be calculated as an example by embedding the aforementioned functions to the analyzer. A parabolic equation is applied for the approximate equation in this research.

A real tool wear data (Anzai, 2003) is used in order to confirm the tendency depicted in Fig.6. For this case study, cutting tool is a ball end mill with R10 and 2-flute, and workpiece is PX5. Cutting speed is varied from 50 to 550m/min, the axial depth is 0.5 mm, the radial depth is 0.8 mm, the feed per tooth is 0.15 mm/tooth and the cutting length is 56.25m. The coolant is also used for this cutting. Figure 7 describes the relation of tool wears according to cutting speed. Here, a flank wear is used to distinguish its tool life and then the threshold of maximum tool wear is assumed to be 0.8 mm, and the tool life in time domain is obtained.

Figure 8 shows a relation of equivalent CO_2 emission according to the cutting speeds. The approximate equation is obtained regarding the plotted data as follows.

$$y = 3.38 \times 10^{-2} x^2 - 27.0x + 6.02 \times 10^3 \qquad (10)$$

Where, y means equivalent CO_2 emission and x means cutting speed. The minimum cutting speed is obtained by the iterative calculation and becomes 398.9 m/min (about 12702 rpm).

A research introduces an importance about the decision of an optimum cutting condition achieving low environmental burden with using virtual reality technology before a real machining operation (Shao, 2010) based on my previous research (Narita, 2009), but any concrete ways to decide the optimum cutting condition from the view point of the environmental burden haven't been proposed so far. This is the first proposition how to decide the optimum cutting conditon achieving low environmental burden as show in this example. I believe the feasibility of the environmental burden analyzer can be described in this paper.

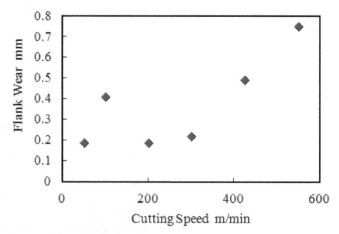

Fig. 7. Tool wears according to cutting speed (Anzai, 2003)

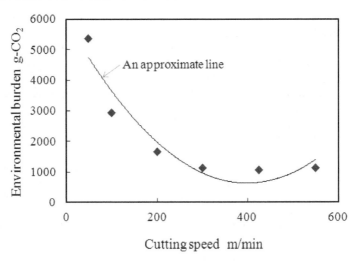

Fig. 8. Environmental burden vs. cutting speed

6. Conclusion

Conclusions are summarized as follows:

1. A algorithm to calculate environmental burden due to machine tool operations was proposed and the environmental burden analyzer for machine tool operations was developed.
2. A decision method of cutting conditions to achieve minimum environmental burden with using the developed analyzer was also proposed.
3. The feasibility of the environmental burden analyzer and the decision method of cutting conditions to achieve minimum environmental burden were demonstrated through examples.

7. Acknowledgment

I would like to express my sincere appreciation to the stuff of OKUMA Corp. for thoughtful support.

8. References

Anzai, M., The Cutting Characteristic of Mold Steel Under The High-Speed Milling Condition, *Nachi Technical Report*, Vol. 11 A1, 2006 (in Japanese).

Bureau of Waterworks Tokyo Metropolitan Government, (2003), Environmental report of Tokyo waterworks 2002(in Japanese).

Diaz, N., Choi, S., Helu, M., Chen, Y., Jayanathan, S., Yasui, Y., Kong, D., Pavanaskar, S. and Dornfeld, D., , (2010), Machine Tool Design and Operation Strategies for Green Manufacturing, *Proceedings of 4th CIRP International Conference on High Performance Cutting* , Gifu, Japan, pp.271-276.

Fujimura, Y., Yasui, T. , (1994), Machine tool and manufacturing system, KYORITSU SHUPPAN Co. Ltd., (in Japanese).

Hara, Y., Tanaka, T. and Saito, Y. , (2005), Evaluation Method of the Power Consumption for Machine Tool, Proceeding of the 2005 JSPE Autumn Annual Conference (in Japanese) , Kyoto, Japan, pp.744-745.

IPCC Fourth Assessment Report (AR4) , (2007), Changes in Atmospheric Constituents and in Radiative Forcing.

Jiang, B., Qi, C., Xia, D., Cao, S., and Chen B. , (2011), Dynamics Characteristic of Cutter Wear in High Speed Milling Hardened Steel, Advanced Science Letters Vol. 4, 3103-3107

Mizukami, H., Yamaguchi, R., Nakayama, T., Maki, T., , (2002), Off-gas Treatment Technology of ECOARC, *NKK Technical Report (in Japanese)*, No.176, pp.1-5.

Nansai, K., Moriguchi, Y. and Tohno, S. , (2002), Embodied Energy and Emission Intensity Data for Japan Using Input-Output Tables (3EID)–Inventory Data for LCA-, Center for Global Environment Research, National Institute of Environmental Studies, Japan.

Narita, H., Chen, L.Y., Fujimoto, H., Shirase, K. and Arai, E. , (2006), Trial-Less Machining Using Virtual Machining Simulator for Ball End Mill Operation, *International Journal of the Japan Society of Mechanical Engineers*, Series C, Vol.49, No.1, pp. 50-55.

Narita, H. and Fujimoto, H., Analysis of Environmental Impact Due to Machine Tool Operation, International Journal of Automation Technology, Vol.3, No.1, pp.49-55, 2009.

Organisation for Economic Co-operation and Development (OECD), (2009), Sustainable Manufacturing and Eco-innovation: Towards a Green Economy, OECD June 2009 Policy Brief.

Osaka prefecture, (2003), guideline index of waste material incinerator in Osaka prefecture, Osaka prefectural announcement No.618 (in Japanese).

Osaka prefecture, (2005), Osaka prefectural statistical yearbook 2004 (in Japanese).

SETAC, (1993), Guidelines for Life-Cycle Assessment: A Code of Practice.

Shao, G., Kibira, D. and Lyons, K., (2010), A Virtual Machining Model For Sustainability Analysis, *Proceedings of ASME 2010 International Design Engineering Technical Conference & Computers and Information in Engineering Conference*, DETC2010-28743, August 15-18, Montreal, Quebec, Canada.

Sheng, P., Bennet, D., Thurwachter, S., / B.F. von Turkovich, (1998), Environmental-Based Systems Planning for Machining, *Annals of the CIRP*, Vol.47/1, p.409-414.

Shimoda, M. , (2000), LCA case of machine tool, *Symposium of 2002 Japan Society for Precision Engineering Spring Annual Meeting "Leading-Edge Trend of Environmental Impact Evaluation for Inverse Type Design and Manufacturing"*, Nagoya, Japan, p. 37-41, (in Japanese).

Tokyo Electric Power Company, (2005), The Earth, People & Energy TEPCO Sustainability Report 2004

Touma, S., Ohmori, S., Kokubo, K., Tateno, M. , (2003), Evaluation of Environmental Burden in Eco-friendly Machining Method using Life Cycle Assessment Method – Estimation of Carbon Dioxide Emission in Eco-friendly Turing Method– *Journal of the Japan Society for Precision Engineering*, (in Japanese), Vol.69, No.6, p.825-830.

Modelling and Implementation of Supervisory Control Systems Using State Machines with Outputs

Moacyr Carlos Possan Junior[1,2] and André Bittencourt Leal[2]
[1]Whirlpool Latin America
[2]Santa Catarina State University – UDESC
Brazil

1. Introduction

The growth of the complexity of automated systems in industry has occurred extensively in recent years due to the production demand increase, quality improvements and flexibility to restructure the manufacturing systems in order to satisfy new procedures. Nevertheless, the evolution of control devices and their functionalities, such as processor speed, memory and network communication has advanced in parallel with the factories' requirements. Although the evolution of automation in industrial processes is a fact, there is still a scarcity of formal methods for analysis, project and implementation of control systems for Discrete Event Systems (DES) in order to reduce the development time, reduce human resources investment, and satisfy the operational requirements for certain systems in an effective way. Furthermore, the occurrence of programming bugs resulting in errors due to interruption in the process and losses due to poorly designed software is obviously unacceptable nowadays in this market that has a just-in-time mindset and is strongly focused on profits.

Usually, the projects for supervisory control systems are based on the knowledge of the system's practitioner, according to his experience in programming. The usage of formal methods is sparse, such that the reuse of documentation and source code as well as the dissemination of the knowledge generated are both impaired. Moreover, the automation of manufacturing systems has brought an increase in the complexity of control systems, so that to elaborate and implement robust and reliable control logic is not a trivial task. In order to minimize the risks due to programming errors and to permit a formal method for modelling DES, Ramadge & Wonham (1989) introduced the Supervisory Control Theory (SCT), which guarantees optimal control logic (nonblocking and minimally restrictive) for these systems.

A Discrete Event System (DES) consists of a system with discrete states that are driven by events. In other words, its state evolution depends on the occurrence of discrete asynchronous events over time (Cassandras & Lafortune, 2008). Discrete Event Systems are quite common in the industry nowadays and the events may be classified as uncontrollable and controllable. Examples of controllable events are the start and end of an operation and examples of uncontrollable events are the activation and deactivation of a presence sensor.

The Supervisory Control Theory (SCT) is already widespread in the academic environment, using the automata theory as a base to model the control systems. However, such a theory is not common in the industrial environment. Therefore the resolution of supervisory control problems has been done without the usage of a formal procedure.

The SCT allows the solution of control problems in a systematic manner. This technique guarantees that the resulting supervisor will satisfy the specifications imposed by the designer, avoiding general control issues, such as blocking. Besides, due to its heuristic nature, the SCT facilitates the code writing changes before implementation in a controller, in case there is some inclusion/exclusion of devices or changes in the system layout.

Although the SCT provides an automatic method to synthesize control systems for DES, when analysing the monolithic supervisor obtained from the SCT, it is difficult to visualize the process dynamics in an easy way as the system complexity grows. That occurs due to the large number of states and no distinction about what kind of event, uncontrollable or controllable, has priority to occur when the supervisor is in a certain state. Furthermore, the implementation of this supervisor in a controller will require considerable non-volatile memory, neither being an elegant solution nor justifying the adoption of a formal method. PLC (Programmable Logic Controller) implementation of DES supervisory control was discussed in many works, as in (Ariñez et al., 1993; Lauzon, 1995; Leduc & Wonham, 1995; Leduc, 1996; Lauzon et al. 1997; Fabian & Hellgren, 1998; Dietrich et al., 2002; Hellgren et al., 2002; Liu & Darabi, 2002; Music & Matko, 2002; Queiroz & Cury, 2002; Chandra et al., 2003; Hasdemir et al., 2004; Manesis & Akantziotis, 2005; Vieira et al., 2006, Noorbakhsh & Afzalian 2007; Hasdemir et al., 2008; Silva et al., 2008; Leal et al., 2009; Uzam et al., 2009; Moura & Guedes, 2010).

In brief, other works presented methodologies using extended automata with variables in an attempt to minimize the exponential growth of states resulting from the automata composition, such as (Chen & Lin, 2000; Yang & Gohari, 2005; Gaudin & Deussen, 2007; Skoldstam et al., 2008), amongst others.

This chapter intends to propose a formal methodology to model control systems for industrial plants through extended automata called Mealy state machines (Mealy, 1955) and subsequent implementation in Programmable Logic Controllers (PLCs) using the Ladder language. An algorithm proposed by Possan (2009), which explores the benefits of SCT to design a supervisor, is used to convert the automaton which represents the supervisor to a finite state machine with outputs. Another algorithm is then used to reduce the state machine to have a simplified structure implemented in a PLC. The simplified state machine representation is a different way of viewing a DES, resulting in a systematic way to implement the source code in a controller with reduced memory usage. The code implementation takes into consideration some common issues found in synchronous controllers, such as PLCs.

A Mealy state machine is a finite state machine with outputs, composed of an oriented graph where the nodes are called states and arcs are called transitions. It is a powerful model to represent the behavior of processes in general.

The chapter is structured as follows: Section 2 introduces the proposed methodology; Section 3 covers the monolithic approach defined by the SCT; Section 4 shows a system to be used as example to illustrate the modelling and implementation; Section 5 presents in detail the transformation algorithm to obtain a Mealy state machine; Section 6 relates to the state machine simplification procedure; Section 7 presents common issues found during supervisor implementation in synchronous controllers while Section 8 describes how to implement the simplified state machine in PLC using the Ladder language. Finally, Section 9 covers the conclusion.

2. Methodology for the control system design

Figure 1 shows an overview of the proposed methodology. It starts with the supervisor synthesis. The synthesis is done based on the SCT and a monolithic supervisor is obtained. The supervisor is then used as input for the transformation algorithm to obtain the state machine. The machine is then simplified to have a reduced number of state transitions. The simplified machine, on the other hand, represents a model to generate the code for a controller (PLC, microcontroller or any other data processing unit). This chapter is focused on the state machine implementation using the Ladder language for PLCs.

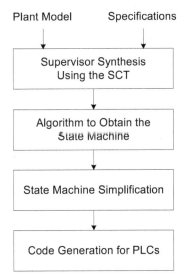

Fig. 1. Proposed methodology

The synthesis of the supervisor and the obtaining and simplification of the state machine will be described in the next three sections.

3. Supervisor synthesis using the SCT

The monolithic approach for the synthesis of an optimal supervisor (nonblocking and minimally restrictive) is based on three main steps:

a. obtain a model for the physical system (plant) to be controlled;
b. obtain a model which represents the specifications to be respected; and
c. synthesize a nonblocking and optimal control logic.

The plant model is built through the synchronous composition (Cassandras & Lafortune, 2008) of all the existing subsystem models in the system. The same procedure is done to build the specification model. The plant and the target language, obtained through the synchronous composition of the plant with the specification, will be used as input to obtain the monolithic supervisor. This procedure can be done using the computational tool named *Grail for Supervisory Control* (Reiser et al., 2006).

In the SCT, the plant is assumed to spontaneously generate events. The supervisor observes the string of events generated by the plant and might prevent the plant from generating a subset of the controllable events, thus disabling them. However, the supervisor has no means of forcing the plant to generate an event, as shown in Figure 2-a.

In practice, the modelled behavior of the plant does not correspond exactly to the real behavior due to the assumption that controllable events are not generated by the plant, as presumed by the SCT. This is because in most real systems, the events modelled as controllable correspond to commands that actually must be generated by the control system. These commands must be sent by the controller to the actuators because they would not occur spontaneously. Thus, the implementation is performed according to the structure proposed by Queiroz & Cury (2000) to keep such coherence. Figure 2-b shows the representation for real systems, where the electric signals coming from the sensors (responses from the plant) correspond to the observed (uncontrollable) events while the electric signals sent to the actuators (actions in the plant) correspond to the disabled (controllable) events.

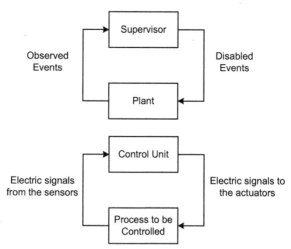

Fig. 2. Monolithic approach: (a) Ramadge-Wonham Framework, (b) Similar representation for a real system

4. A motivation example

In this section, a manufacturing system is used to demonstrate the proposed methodology. This system is composed of three apparatus and two intermediary buffers with a capacity of one which are available between the apparatus, as illustrated in Figure 3.

The apparatus are represented by Ai, where $i = 1, 2, 3$ and the buffers are represented by Bj, where $j = 1, 2$.

Fig. 3. Manufacturing system

The controllable events that correspond to the start of the apparatus' operation are represented by a_x, while the uncontrollable events that correspond to the end of operation are represented by b_x, where $x = 1, 2, 3$.

The plant and specifications are modelled using automata and the controllable events are represented with a dash. The behavior of each apparatus (or subsystem) can be modelled by the automaton shown in Figure 4. Notice that state 0 is double circled. This is a marked state that represents a completed task.

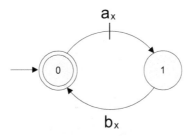

Fig. 4. Automaton for the apparatus Ax

When modelling this plant, only the apparatus were taken into consideration. The buffers were considered only in the in the control specification model.

The specifications for this system are restrictions of coordination to avoid overflow (the apparatus finishes its task but the output buffer is already full) or underflow (the apparatus starts working without any item to fetch from the input buffer). Those restrictions point out the idea that it is necessary to alternate b_1-a_2 and b_2-a_3, respectively. This means that the start of operation for an apparatus (event a_{x+1}) will only be allowed when its input buffer is loaded (event b_x), as shown in Figure 5.

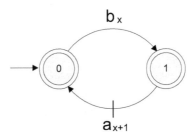

Fig. 5. Automaton for the control specifications

The synchronous composition of the plants with the specifications will result in the target language (Ramadge & Wonham, 1989). The calculation of the minimally restrictive and nonblocking supervisor is based on an iterative process which identifies and eliminates bad states in the automata that models the target language.

The monolithic supervisor found with the usage of *Grail* (Reiser et al., 2006) has 18 states and 32 state transitions.

The supervisor can be reduced for a later comparison with the state machine. This procedure is done to simplify the supervisor size and also the number of transitions (Su & Wonham, 2004). An algorithm for supervisor reduction is used to obtain fewer states and transitions than the original supervisor (Sivolella, 2005).

Figure 6 shows the automata which represents the reduced supervisor. The disabled events for each state are represented by red dashes.

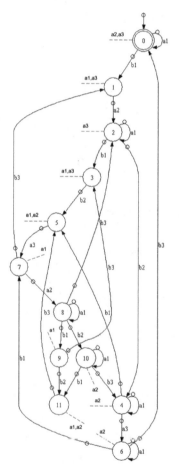

Fig. 6. Reduced supervisor and disabled events for the manufacturing system

Notice that the reduced supervisor has 12 states and 20 state transitions. The supervisor reduction algorithm creates self-loop transitions. However, these transitions are not relevant because they do not cause a state change.

Transitions among the states can occur either by controllable events (a_x) or by uncontrollable events (b_x). In case the designer intends to implement the code in the controller based on this model, he will have to decide on the kind of event to give priority. Furthermore, it is not trivial to visualize what are the active apparatus and what events modelled as uncontrollable are expected to occur at a certain state of the supervisor.

5. Algorithm to obtain the state machine with outputs

The state machine obtained with the proposed algorithm consists of a Mealy machine (Mealy, 1955). In a Mealy machine, a transition can have one or more output actions (set one or more controllable events) and any output action can be used in more than one transition. The output actions are not associated with the states, which are passive. Thus, the actions can be associated with more than one state.

A simple example of a Mealy machine is shown in Figure 7. Transitions and actions are separated by dashes. It is a machine with two states where, when in State 1, Transition 1 makes the machine to go from State 1 to State 2 and takes Action 1. When in State 2, Transition 1 plus Transition 2 makes the machine go from State 2 to State 1 and Action 2 is taken.

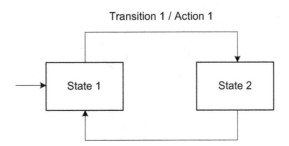

Transition 1 / Action 1

State 1   State 2

Transition 1, Transition 2 / Action 2

Fig. 7. Example of a Mealy machine

In this proposal, the transition between two states is performed by means of one or more uncontrollable events in the system. For each transition, an output action may be generated.

The algorithm proposed to create the state machine works in an iterative process, looping through the input data to obtain the states, transitions and actions that compose a finite state machine. As soon as all the input data is processed, the state machine is completed. The algorithm is shown in Figure 8.

Input data for the algorithm is the information about the plant, the supervisor and the list of disabled events. Output data are the states, transitions and actions which compose the state machine.

In the transformation process from the supervisor to the state machine, the output actions correspond to the controllable events in SCT while the transitions among the states correspond to the uncontrollable events.

The initialization considers the initial states of the input data. These data compose the starting point of the state machine representing the condition where the physical process has not started.

The next step is to create a states queue. The queue is required to store the states that are being obtained iteratively to be treated after the treatment of a current state has finished. The queue consists of a First In, First Out (FIFO) structure. A while loop is suggested to treat all the states available until the states queue becomes empty.

```
1: Read the supervisor (S) automata and the disabling map
(DM)
2: Split the supervisor automata according to its events
controllability – Su (uncontrollable) and Sc (controllable)
3: Create the initial state for the machine = initial state of
supervisor automata
4: Create a list with the next states which are obtained from
the current states that need to be processed
5: while (States List > 0) do
6:          Read state from the list and consider as a current
state
7:          Create a Transitions List for this state
8:          for (Transitions List) do
9:                    Read Su, MD, Sc
10:                   if (Sc evolved) then
11:                             Update MD e Sc
12:                   else
13:                             Create a new next state
14:                   end if
15:                   if (State was already created) then
16:                             do nothing
17:                   else
18:                             Add the next state to the States
                                List according to the next state
                                from the supervisor automata
19:                   end if
20:          end for
21:          Decrement States List
22: end while
23: Remove unreachable states
24: Save the Monolithic Mealy State Machine
```

Fig. 8. Algorithm to obtain the state machine

For each state, a transition queue is created as well. A "for loop" is suggested to treat all the valid transitions for that specific state until all the transitions available in the queue are processed.

In order to create the list of valid transitions for a certain state in the state machine, it is first necessary to divide the plant and the supervisor in two parts according to the controllability of the events and their transitions. Thus G_u and S_u will have the list of automata state transitions due to uncontrollable events. The same happens for the controllable part.

The data reading sequence begins at the uncontrollable part (G_u and S_u). After that, the algorithm evaluates the disabled events for the actual state of the supervisor. This will define if the controllable part of the system (G_c and S_c) can evolve or not.

In other words, the resulting state machine gives priority to the occurrence of uncontrollable events, waiting to receive some response from the physical system to make a decision about what controllable events to disable.

The valid transitions for a certain state are the uncontrollable events that create state evolutions from the current active states in the uncontrollable part of the plant and supervisor. The combination of more than one uncontrollable event is also considered as a transition.

After the uncontrollable part is processed, the controllable part is processed. The controllable disabled events are evaluated according to the current state the supervisor is in. While the disabled events are forbidden to occur, the remaining ones may originate the actions.

The valid actions for a certain state are the controllable events that are not disabled at the supervisor state and cause states evolution in the controllable part of the plant and the supervisor.

When an action occurs, the algorithm checks if the controllable part of the supervisor has evolved. If so, then the disabled events for the destination state are evaluated in order to verify if another action may occur. This step assures that all the actions possible to occur for the same transition are processed. It means that more than one action can occur for the same transition.

If S_c has not evolved, then a new state is created. This state is compared with the other states and if it already exists, it is ignored. Otherwise, it is added to the states queue to be treated later.

For each transition due to the uncontrollable events, G_u and S_u obtained from the SCT evolve. The same happens for the controllable part in the occurrence of an action.

This methodology is consistent with the Mealy machine approach, where the outputs (actions) depend on the current state and valid inputs (transitions).

After the procedure for creating a transition and corresponding actions is finished, the algorithm will treat the remaining transitions available in the transitions queue. When all the transitions in the queue are treated, the algorithm will evaluate the next state available in the states queue. These iterative processes are performed until the states queue is empty. This means that the finite state machine has been completed.

The state machine for the manufacturing system with the proposed algorithm is shown in Figure 9, composed of 8 states and 22 state transitions.

The states are named according to the apparatus that are operating at a certain point in time and the buffers that are full. The transitions are represented by the uncontrollable events, and the taken actions, if any, are separated from the transitions by a slash (/). The disabled events are represented by red dashes. Notice that in this model the transitions may occur due to more than one uncontrollable event, and the actions, if any, may be due to more than one controllable event.

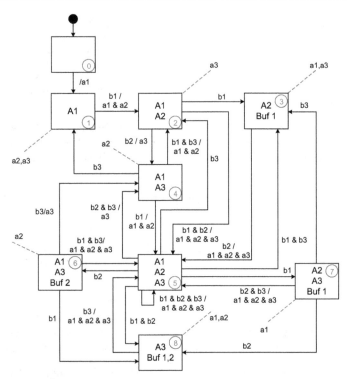

Fig. 9. Mealy state machine

6. State machine simplification

When the state machine model considers the treatment of more than one uncontrollable event, this may result in the disadvantage of an exponential growth of transitions, depending on the number of plants modelled and how many may be enabled at the same time. The number of transitions created for a certain state is on the order of $2^n - 1$, where n represents the number of uncontrollable events present in the model and possible to occur for that state. Therefore, for larger and more complex systems, the code size would be affected significantly to satisfy that condition, being a convincing reason for not using such a methodology. An alternative solution for that is to consider a reduced state machine where the state transitions are represented only by transitions that result in actions. Transitions that do not result in actions are represented by self-loops inside their current state. For example, for state 2 of the state machine represented in Figure 9, transition b_1 could be represented as a self-loop. Although the state machine evolves to state 3, there is no action taken during this transition.

Although they do not result in actions, these transitions are important to represent the plant dynamics and cannot be simply disregarded during the implementation process. The controller program must capture their occurrence and store it in some internal variable to be used in the decision-making process when other transitions occur. The algorithm presented in Figure 10 describes the process to reduce a monolithic Mealy machine.

```
1: Read the Mealy machine
2: for State = from i to n do % where n is the number of states
3:        for Transition = from i to t do % where t is the
number of transitions from the current state
4:                if (Transition results in action) then
5:                Keep the transition in the reduced machine
6:                else
7:                Create a self-loop represented by dashed
                  lines in the current state
8:                end if
9:        end for
10: end for
11: Remove unreachable states
12: Save the Reduced Monolithic Mealy State Machine
```

Fig. 10. Algorithm to reduce the state machine

It is important to emphasize that such a procedure to reduce the Mealy machine does not necessarily result in a minimal state machine.

Figure 11 shows the reduced state machine for the manufacturing system. This state machine contains only four states and seven transitions. The transitions illustrated by solid lines represent the occurrence of an action, regardless of whether the transition occurred from one state to another. The transitions illustrated by self-loops in dashed lines inside a current state represent that, although a transition has occurred, an action is not fired.

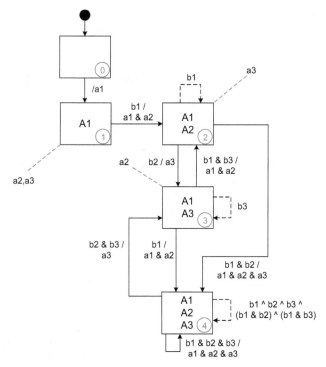

Fig. 11. Reduced Mealy state machine

Consider for instance state 2 of the reduced state machine. If transition b_2 occurs, the state machine evolves to state 3. In the case that transition b_1 occurs, the model illustrates that situation as a self-loop represented by dashed lines, which means that the practical implementation in the controller must guarantee the proper storage of this information in some internal variable.

When event b_2 occurs with transition b_1 already enabled, transition b_1 & b_2 will be activated, so that the state machine evolves to state 4 and actions a_1, a_2 and a_3 are taken.

The disabled actions shall be represented in the corresponding states to illustrate the control actions forbidden to occur. The states which became unreachable if compared to the original state machine are eliminated by this reduced model.

Transitions due to more than one uncontrollable event and action due to more than one controllable event are still represented in this model. That is relevant information which helps the designer when he intends to implement the control system, so that the program allows several events to be executed inside the same scan cycle of the synchronous controller.

In addition, a new logic operator appears in this model. The operator "$\wedge$" represents an *exclusive or* condition. In order to understand its function, consider state 4 of the reduced state machine, for example. If any of the transitions listed in the self-loop with dashed lines occur, namely, $b_1 \wedge b_2 \wedge b_3 \wedge (b_1$ & $b_2) \wedge (b_1$ & $b_3)$, none of the actions will be taken and those transitions will remain enabled. Actions will be taken only when the transitions b_2 & b_3 or b_1 & b_2 & b_3 become valid. The operator "$\wedge$" appears only for the states with a self-loop represented by dashed lines.

7. Issues with Implementing supervisors in synchronous controllers

According to (Fabian & Hellgren, 1998), "the supervisor implementation is basically a matter of making the Programmable Logic Controller (PLC) behave as a state machine". However, that is not a simple task. Certain issues appear during the implementation process in synchronous controllers, such as PLCs and computers. Those problems exist regardless of the model used to represent the supervisors, by means of automata, Petri nets or colored Petri nets (Basile & Chiacchio, 2007). Such problems are explored in the structure presented here and a solution is presented where possible, in accordance with the Ladder language definition described by the IEC-61131-3 (1993) standard.

7.1 Causality

The SCT considers that all events are generated spontaneously by the plant and that the supervisor tracks the sequence of events generated by the plant and also acts to disable the controllable events in order to avoid any infringement of the control specifications. However, in most practical applications, the controllable events are not generated spontaneously by the physical plant, but only the feedback due to sent commands. So, the question "who generates what?" must be answered (Fabian & Hellgren, 1998), or in other words, who is responsible for the generation of certain kind of events, the supervisor or the plant? (Vieira, 2007).

The supervision scheme proposed by the SCT was already illustrated in Figure 2-a. In this structure, the plant is the responsible for events generation, both controllable and uncontrollable, while the supervisor is just an observer and disables a set of controllable events to satisfy control requirements.

However, in most practical applications, the events modelled as controllable correspond to commands that, indeed, shall be generated by the PLC and sent to the actuators, because they do not occur spontaneously. The plant generates only the uncontrollable events, as a feedback to the stimulus sent by the PLC through the controllable events. Figure 2-b describes the control structure usually employed in practice (Malik, 2002).

In this chapter, the causality problem is solved by representing the model using state machines with outputs instead of automata. Therefore, when this transformation is performed, the SCT changes to a model with inputs and outputs, as suggested by Malik (2002). The signals originated by the sensors correspond to the uncontrollable events while the signals generated by the actuators correspond to the controllable events.

7.2 Event detection

The synchronous nature of PLCs may create issues during the detection of uncontrollable events as described below.

7.2.1 Signals and events

Some implementation problems are due to the lack of an easy way to translate supervisors based on discrete event systems, which are symbolic, asynchronous and occur in discrete instants of time in a synchronous universe and are based on signals such as those from the PLC (Fabian & Hellgren, 1998). In order to avoid a discrepancy between the theory and practice, a signal cannot generate more than one uncontrollable event during an operational cycle of the PLC (Basile & Chiacchio, 2007).

7.2.2 Avalanche effect

The PLC signals may assume boolean values and are sampled periodically. Thus, in order to implement supervisors according to the SCT in PLCs, the events are associated with changes in the PLC input signals, which may cause what is called the *avalanche effect*. This effect occurs when a value change in an input signal is registered as an uncontrollable event and makes the software jump to an arbitrary number of states during the same scan cycle of the PLC. This may occur specifically if a certain uncontrollable event is used to trigger several state transitions in a list, creating behavior similar to an avalanche.

Figure 12 shows an example of the avalanche effect for a conventional supervisor implementation in PLC. The supervisor transitions from state 0 to state 2 with the occurrence of event b_1. It should transition from state 1 to state 2 only with the occurrence of a new event b_1, or transition from state 1 to state 3 with the occurrence of an event b_2. However, the implementation proposed on the right side of Figure 12 does not restrict that event b_1 permits the transition from state 1 to state 2 in the same scan cycle of the PLC. So, the avalanche effect introduces a transition directly from state 0 to state 2, so that, even if event b_2 happened, it would not have any effect on the controller dynamics. This is clearly unsatisfactory.

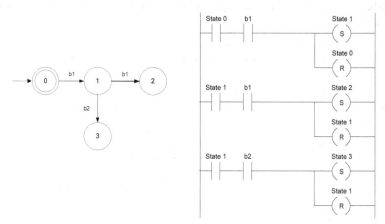

Fig. 12. Example of the Avalanche Effect

In order to solve this problem, Vieira (2007) proposes a procedure where a variable is associated with all the transitions available for the supervisor. That variable is activated every time a new state transition occurs, being deactivated at the end of the section of source code where the supervisor structure is implemented. Every state transition includes the negation of this variable as a condition. Such a solution solves the problem, but has the disadvantage of allowing the occurrence of only one uncontrollable event inside the same PLC scan cycle.

The solution proposed in this chapter consists of the deactivation of the variable corresponding to the event which occurred as soon as a state transition due to that event happens. In this way, the supervisor deactivates that specific event which occurred, allowing transitions due to other events available in the model still to happen. During a scan cycle, if events related to asynchronous specifications occur, they will be treated in this same cycle. Such a proposal solves the avalanche effect problem and permits several uncontrollable events to be treated inside the same PLC scan cycle. Moreover, the constraint that more than one event is treated in the same scan cycle results in another problem called *loss of information* (Vieira, 2007). Such a problem occurs in situations where several events happen and only one is treated per scan cycle. Then, the information related to the occurrence of the remaining events is lost. That issue does not occur in the methodology proposed here due to the possibility of treating all other events which occurred in the same scan cycle. This is possible because there is not just one general variable which is activated when a state transition occurs, but a specific treatment for each event, where the other events remain enabled to fire transitions.

7.2.3 Inability to recognize the order of events

This simultaneity problem was covered first by Fabian & Hellgren (1998) considering the implementation of supervisors in PLCs. When two or more input signals change their values between two input readings, those changes will be stored as a simultaneous change of uncontrollable events, so that it is not possible to identify which one occurred first, because the PLC performs a synchronous reading of its inputs.

It is noticeable that signal changes may be simultaneous or not, but the events will always be stored as simultaneous during the reading. That problem is called *simultaneity* (Fabian & Hellgren, 1998). Thus, in order to avoid implementation problems, the supervisor shall have a control action which does not depend on the different interleaving of the uncontrollable events. Fabian & Hellgren (1998) define such a property as *interleave insensitivity* of uncontrollable events. These authors proposed an algorithm to detect if a supervisor is interleaving insensitive. If not, the order in which the uncontrollable events occur cannot be recognized if the controller utilized is synchronous.

8. Code generation for PLCs

The scan cycle of a program in the PLC follows the sequence: input read, control logic execution and output write. That synchronous behavior of the PLC forces the outputs to be updated only at the end of the scan cycle. Due to that, the activation of the actuators requires a specific treatment. Looking at the state machine's structure, it may happen that as soon as it finishes the operation, an apparatus may be requested to start a new operation cycle inside the same scan cycle in which the previous operation had just been finished. If this happens during a scan cycle, that variable would assume a low logic level (end of operation) and return to a high logic level (start of operation). However, as the PLC writes the values stored in its internal memory to the physical outputs only when its execution cycle is finished, it does not recognize the process of operation end/start, so that its physical output will be kept active all the time during the scan cycle. In order to avoid that, variables are added to represent the evolving of the plants and guarantee the proper synchronism during the system dynamics. Those variables are called *Plant i*, with *i* varying from 1 to *n*, where *n* is the number of plants modelled in the global system. That variable is enabled every time an apparatus finishes its operation. This procedure guarantees that the apparatus is not requested to start operating again inside the same scan cycle when its operation end is detected. It will be turned on again only in the next scan cycle.

As in the state machine model, for the implementation, the variables b_i represent the uncontrollable events (transitions) while variables a_i (actions) represent the controllable events.

The Ladder code can be split into five blocks, called by a main organizational block in the following order: initialization, inputs, transitions/actions, disabling and outputs. They implement the reduced state machine shown in Figure 11.

8.1 Initialization

The initialization starts the state machine and puts it into its initial state, and enables the controllable event a_1 in order to start the process, as shown in Figure 13. Other variables, such as the ones responsible for uncontrollable events *bi* and the evolution of plants *Plant i*, are disabled.

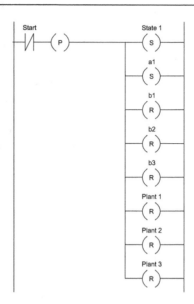

Fig. 13. Initialization

8.2 Inputs

The variables responsible for transitions b_i will be activated in the controller only when an edge rise occurs (from logic level 0 to 1) in the corresponding inputs. Therefore, a pulse detector is required for each input in order to signal the corresponding event, as shown in Figure 14. The variables related to the uncontrollable events are activated in this block. During the program execution, if they result in some action, they will be disabled. If they result only in self-loops where no action is taken, as represented by dashed lines in the state machine, these variables remain active until the moment that some transition which results in an action occurs later.

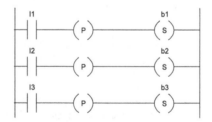

Fig. 14. Inputs

8.3 Transitions/actions

The requirement for a transition to occur is that the state machine must be in a certain state and one or more uncontrollable events that result in some action happens. If these requirements are satisfied, then the next state is activated and the current one is deactivated. The uncontrollable events responsible for the transition are deactivated to avoid occurrence of the avalanche effect (Fabian & Hellgren, 1998). When an uncontrollable event happens for a certain plant, the variable corresponding to that plant, *Plant i*, is activated in order to avoid that the action corresponding to the start of the operation for that apparatus occurs during the same scan cycle, being treated only in the next scan cycle. This is due to the specific treatment required for the actions, as described previously. The actions, if any, will be activated to allow the corresponding plant to evolve in the same scan cycle or only in the next scan cycle if forbidden to evolve in the current cycle. Figure 15 shows this block for the manufacturing system. Verify the representation of the reduced state machine shown in Figure 11 in order to compare the theoretical model with the practical one. Consider for instance that the state machine is in state 3. Here, two transitions are possible: due to both events b_1 and b_3 or only due to event b_1. The transition b_1 & b_3 (due to events b_1 and b_3) must always appear first in the Ladder diagram due to this problem of simultaneity (Fabian & Hellgren, 1998). This is because in the case that events b_1 and b_3 occur and the transition only due to event b_1 is implemented first, the latter will be executed in the program and the variable b_1 will be deactivated. In this way, the transition due to events b_1 and b_3 will never happen in practice.

Consider now that the state machine is in state 4 and the transition b_1 & b_2 & b_3 occurs. The state machine remains in the same state and, as all the apparatus finish their operation, no action will be taken in the current scan cycle. However, as the actions a_1, a_2 and a_3 are activated, in the next scan cycle all the apparatus will be turned on again. Thus, such a methodology guarantees that it is necessary to wait only one scan cycle for the enabled actions to be taken.

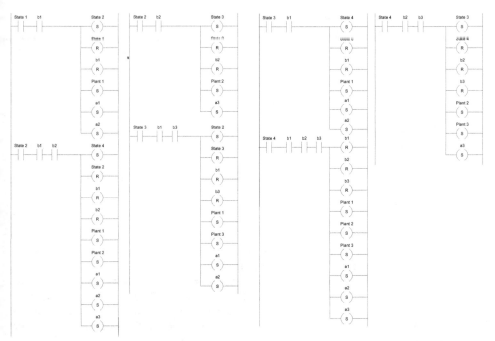

Fig. 15. Transitions/Actions

8.4 Disabling

This block is responsible for disabling the control actions in the machine states. This means that if the state machine reaches a certain state, the actions related to this state are forbidden to occur, being disabled. Figure 16 shows the disabling block for the manufacturing system. The first two rungs on the Ladder code represent the states where the corresponding control actions are disabled. Furthermore, the disabling may occur if some uncontrollable event is signalled but it does not allow an action to be taken (represented by a self-loop with dashed lines in the reduced state machine), so that the start of operation in the corresponding plant is forbidden. In the Ladder diagram, it is enough to disable the action related to a certain *Plant i* when the uncontrollable event *bi* occurs. In other words, when b_1 occurs, it is enough to disable *Plant 1*. When b_2 occurs, it is enough to disable *Plant 2*, and so on. This is because although the event *bi* occurred, it did not effectively generate a transition in the state machine, and therefore, the plant was not disabled, as shown in the remaining rungs of the Ladder diagram.

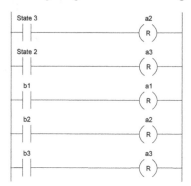

Fig. 16. Disabling

8.5 Outputs

A physical output is activated only if the action related to a controllable event is taken and its corresponding variable, *Plant i*, is not enabled, as shown in Figure 17. If such situation occurs, the coil Q_i which represents the physical output of the PLC will be energized. Yet, at the end of the program, the variables *Plant i* will be disabled in order to return to the initial condition before starting a new scan cycle.

Besides being a simplified implementation model, this solution has the advantage of not restricting more than one uncontrollable event to be treated in the same PLC scan cycle.

In order to have a better understanding of the logic implementation of this supervisory control, consider first that the start of operation is given by the action a_1. After a few PLC scan cycles, the apparatus A1 finishes its operation and the transition due to the uncontrollable event b_1 is generated. According to the reduced state machine, the actions a_2 and a_1 are enabled to occur and the state machine transitions to *state 2*. However, as *Plant 1* finished its operation, only the physical output Q_2 is activated in the same scan cycle. The physical output Q_1 will be activated only in the next scan cycle. Next, the state machine will behave differently depending on which apparatus finishes its operation first. If apparatus

A2 finishes its operation, then transition b_2 will be generated, action a_3 will be taken and the state machine transitions to *state 3*. If apparatus A1 finishes its operation, then transition b_1 will be generated resulting in a self-loop with dashed lines inside *state 2*, because no action is taken. If the PLC identifies the changes in the input signals corresponding to the end of operation in both apparatus A1 and A2 (transitions b_1 and b_2, respectively), then actions a_1, a_2 and a_3 will be taken and the state machine transitions to *state 4*. Similarly, the remaining transitions and actions follow the same dynamics, as illustrated in the reduced state machine model presented in Figure 11.

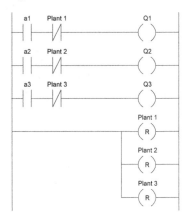

Fig. 17. Outputs

9. Conclusion

The proposed methodology presents a new model to represent supervisors for Discrete Event Systems (DES). The Supervisory Control Theory (SCT) is a convenient methodology to obtain supervisors from simpler models, where automata are useful to represent plants and control specifications so that, starting from simpler models, supervisors may be represented by state machines. In general, the approach based on state machines consists of a model rich in information. It is appropriate to emphasize the importance of the modelling process for the supervisors, because if some control specification changes or a new subsystem is added to the plant, it is necessary to remake the synthesis for the state machine.

The proposed algorithm to transform the automata which represents a supervisor in a state machine allows a reduction in the number of states in the model. That reduced approach can be used as a reference to implement the control system in a data processing unit. It is necessary only to take into account the implementation aspects related to the controller used, based on the constructive aspects of its hardware. In this chapter, solutions were described to avoid the problems usually found when implementing supervisors in synchronous controllers, using a PLC as a target with the source code implemented in the Ladder language.

The example of a manufacturing system demonstrates some aspects related to the optimization in the code size generated resulting in an economy in the non-volatile memory usage and the possibility of treating several events inside the same scan cycle of the controller. For large systems, this approach results in an improvement of the temporal dynamics of the control when several input signal changes and several actions shall be taken in the same scan cycle, ensuring synchronism and minimizing problems due to communication delays.

10. Acknowledgment

The authors thank the financial support provided by CNPq and FAPESC, and also thank Whirlpool Latin America and the UDESC for their support in the realization of this work.

11. References

Ariñez, J.F.; Benhabib, B.; Smith, K.C. & Brandin, B.A. (1993). Design of a PLC-Based Supervisory-Control System for a Manufacturing Workcell, *Proceedings of the Canadian High Technology Show and Conference*, Toronto, 1993.

Basile, F. & Chiacchio, P. (2007). On the Implementation of Supervised Control of Discrete Event Systems. *In: IEEE International Transactions on Control Systems Technology*, Vol. 15, No. 4, pp. 725 – 739, ISSN: 1063-6536

Cassandras, C.G. & Lafortune, S. (2008). *Introduction to Discrete Event Systems*. 2nd Ed. Springer Science, ISBN-13:978-0-387-33332-8, New York, USA.

Chandra, V.; Huang, Z. & Kumar, R. (2003). Automated Control Synthesis for an Assembly Line Using Discrete Event System Control Theory. *IEEE Transactions on Systems, Man, and Cybernetics – Part C: Applications and Reviews*, Vol. 33, No. 2, (May 2003), pp. 284-289, ISSN 1094-6977.

Chen, Y. L. & Lin, F. (2000). Modeling of Discrete Event Systems using Finite State Machines with Parameters. *In: IEEE International Conference of Control Applications*, Anchorage, Alaska, USA.

Fabian, M. & Hellgren, A. (1998). A PLC-based implementation of supervisory control for discrete systems. *In: Proceedings of the 37th IEEE Conference on Decision and Control*, Vol. 3, pp. 3305-3310.

Gaudin, B. & Deussen, P.H. (2007). Supervisory Control on Concurrent Discrete Event Systems with Variables. *In: Proceedings of the IEEE American Control Conference*, pp. 4274-4279, ISSN: 0743-1619, New York, NY, USA.

Harel, D. (1987). Statecharts: A Visual Formalism for Complex Systems. *Science of Computer Programming*, Vol. 8, No. 3 (June 1987), pp. 231-274.

Hasdemir T., Kurtulan, S., & Gören, L. (2004). Implementation of local modular supervisory control for a pneumatic system using PLC. In: *Proceedings of the 7th International Workshop on Discrete Event Systems (WODES'04)*, pp. 27-31, Reims, France.

Hasdemir, T.; Kurtulan, S.; & Gören, L. (2008). An Implementation Methodology for Supervisory Control Theory. *International Journal of Advanced Manufacturing Technology*, Vol. 36, No. 3, (March 2008), pp. 373-385. ISSN 0268-3768.

IEC-61131-3 (1993). International Electrotechnical Commission. Programmable Logic Controllers – Part 3: Programming Languages.

Lauzon, S. C. (1995). *An implementation methodology for the supervisory control of flexible-manufacturing workcells*, M.A. Sc. Thesis, Mechanical Engineering Dept. University of Toronto, Canada.

Lauzon, S. C.; Mills, J. K.; & Benhabib, B. (1997). An Implementation Methodology for the Supervisory Control of Flexible Manufacturing Workcells, *Journal of Manufacturing Systems*, Vol. 16, No. 2, pp. 91-101.

Leal, A. B.; Cruz, D.L.L.; Hounsell, M.S. (2009). Supervisory Control Implementation into Programmable Logic Controllers. In: *Proc. of the 14th IEEE International Conference on Emerging Technologies and Factory Automation*, pp. 899-905, Mallorca, 2009.

Leduc, R. J. (1996). *PLC Implementation of a DES supervisor for a manufacturing testbeb: an implementation perspective*, M.A.Sc. Thesis, Dept. of Electrical and Computer Engineering, Univ. of Toronto, Canada.

Leduc, R. J. & Wonham W. M. (1995). PLC implementation of a DES supervisor for a manufacturing testbed. In: *Proceedings of 33rd Annual Allerton Conference on Communication, Control and Computing*, pp. 519-528, University of Illinois.

Liu, J. & Darabi, H. (2002). Ladder logic implementation of Ramadge-Wonham supervisory controller, *Proceedings of the 6th International Workshop on Discrete Event System (WODES'02)*, pp. 383-389. ISBN: 0-7695-1683-1. Zaragoza, Spain, October 2002.

Malik, P. (2002). Generating Controllers from Discrete-Event Models. *In : Proceedings of the Modelling and Verifying Parallel Processes Movep 02*. pp. 337-342, Nantes, France.

Manesis, S. & Akantziotis, K. (2005). Automated synthesis of ladder automation circuits based on state-diagrams. *Advances in Engineering Software*, 36, pp .225–233.

Mealy, G.H. (1955). A Method to Synthesizing Sequential Circuits. *Bell Systems Technical Journal*. pp. 1045-1079.

Moura, R. & Guedes, L.A. (2010). Control and Plant Modeling for Manufacturing Systems using Basic Statecharts. *In: Programmable Logic Controller*, Guedes, L.A. (Ed), pp. 33-50, INTECH, ISBN 978-953-7619-63-3, Croatia.

Music, G. & Matko, D. (2002). Discrete event control theory applied to PLC Programming, *Automatica*, Vol 43, 1-2, pp. 21-28.

Noorbakhsh, M. & Afzalian, A. (2007). Implementation of supervisory control of DES using PLC. In: *Proc. 15th Iranian Conference on Electrical Engineering (ICEE)*, Tehran, Iran.

Possan, M.C.P. (2009). *Modelling and Implementation of Supervisory Control Systems Based on State Machines with Outputs* (in Portuguese). Thesis (Master Degree), Santa Catarina State University, Joinville, Brazil.

Queiroz, M.H. & Cury, J.E.R. (2000). Modular supervisory control of large scale discrete-event systems. *Discrete Event Systems: Analysis and Control*. Kluwer Academic Publishers *(Proc. WODES 2000)*, Ghent, Belgium, pp. 1-6).

Queiroz, M. H. de & Cury, J. E. R. (2002). Synthesis and implementation of local modular supervisory control for a manufacturing cell, *Proc. of the 6th Int. Workshop on Discrete Event Systems WODES*, pp. 1-6. ISBN: 0-7695-1683-1. Zaragoza, Spain, October 2002.

Ramadge, P.J. & Wonham, W.M. (1989). The control of discrete event systems. *In: Proceedings of IEEE, Special Issue on Discrete Event Dynamics Systems*, Vol. 77, No. 1, pp. 81-98.

Reiser, C.; Da Cunha, A.E.C. & Cury, J.E.R. (2006). The Environment Grail for Supervisory Control of Discrete Event Systems. *In: Proceedings of the 8th International Workshop on Discrete Event Systems*, Michigan, USA, pp. 1-6.

Silva, D. B.; Santos, E. A. P.; Vieira, A. D. & Paula, M. A. B. (2008). Application of the supervisory control theory in the project of a robot-centered, variable routed system controller, *Proceedings of the 13th IEEE International Conference on Emerging Technologies and Factory Automation – ETFA'08*, pp. 1-6.

Sivolella, L. F. (2005). *Contributions for Supervisors Reduction for Discrete Event Systems* (in Portuguese). Thesis (Master Degree), IME, Rio de Janeiro, Brazil.

Skoldstam, M. & Akesson, K. & Fabian, M. (2008). Supervisor Control Applied to Automata Extended with Variables - Revised. Technical Report. Chalmers University of Technology, Gothenburg, Sweden.

Su, R. & Wonham, W. M. (2004). Supervisor reduction for discrete-event systems, *Discrete Event Dynamic Systems*, Vol. 14, No. 1, pp. 31-53.

Uzam, M.; Gelen, G.; & Dalci, R. (2009). A new approach for the ladder logic implementation of Ramadge-Wonham supervisors, *Proc. of the 22nd Int. Symp. on Information., Commun. and Automation Technologies*, pp. 1-7. ISBN 978-1-4244-4220-1, Bosnia 2009.

Vieira, A.D. (2007). *Method to Implement the Control of Discrete Event Systems with the Application of Supervisory Control Theory* (in Portuguese). PhD. Thesis, UFSC, Florianópolis, Brazil.

Vieira, A. D.; Cury, J. E. R. & Queiroz, M. (2006). A Model for PLC Implementation of Supervisory Control of Discrete Event Systems. *Proc. of the IEEE Conf. on Emerging Technologies and Factory Automation*, pp. 225–232. Czech Republic, September 2006.

Yang, Y. & Gohari, P. (2005). Embedded Supervisory Control of Discrete-Event Systems. *In: IEEE International Conference on Automation Science and Engineering*, pp. 410-415, Edmonton, Canada.

3

Digital Manufacturing Supporting Autonomy and Collaboration of Manufacturing Systems

Hasse Nylund and Paul H Andersson
Tampere University of Technology
Finland

1. Introduction

This chapter discusses on the challenges and opportunities of digital manufacturing supporting the decision making in autonomous and collaborative actions of manufacturing companies. The motivation is the change towards more networked collaboration caused by, for example, globally distributed markets and specialization of manufacturing companies to their core competences, their autonomous activities. This situation has led to increasingly complex manufacturing activities in the manufacturing network and the importance of collaboration has become a critical factor. In most cases companies seek to respond to the challenges through cooperation rather than expanding their own operations. The autonomy means that the parties involved in the manufacturing activities do their own tasks by themselves independently from other parties while the collaboration involves the activities that one party cannot do by itself and therefore, co-operation of several parties are required. This kind of situation can be clearly seen in networked manufacturing activities involving several companies, but similarly, inside a company and its one facility, same kind of autonomous and collaborative activities can be recognized. In the discussion, the dimensions of autonomy and collaboration are considered in designing and developing manufacturing systems, as well as in improving the daily operations.

The rest of this Chapter is structured as follows. Section 2 discusses on the main issues behind the research, including Competitive and Sustainable Manufacturing, changeability in manufacturing as well as support from digital manufacturing. In Section 3, a structure for manufacturing systems and entities is proposed, which is the base for the design and development activities of manufacturing systems discussed in Section 4. An academic research environment is introduced in Section 5 describing several of the theoretical aspects discussed before. Section 6 gives a brief conclusion on the topics discussed.

2. Background

The focus of the discussion is on mechanical engineering industry of discrete part manufacturing for business-to-business (B2B) industry, including their part manufacturing and product assemblies. These kinds of products are typically highly customized and tailored to customer needs and requirements with low or medium demand (Lapinleimu, 2001). This type of production usually involves several companies and is formed as a supply network. For example, the production includes a main company, its suppliers and suppliers of suppliers as well as customers and customers of customers. The current manufacturing paradigm, in the above context, has evolved from the early craft manufacturing via mass manufacturing towards mass customization. Typical characteristics that have been recognized include (Andersson, 2007):

- Globally local systems spread over industrial ecosystems and manufacturing networks of their own pros and cons.
- Managing the networked manufacturing, where the importance of procurement and management of knowledge flow increase.
- Specialization to one's core competences and collaborating with others in the manufacturing network.

Early discussions considered whether these characteristics could be fulfilled with developing existing flexible manufacturing systems (FMSs), or to shift to reconfigurable manufacturing systems (RMSs) paradigm. At some point, more ambitious goals were set with the aim to describe a manufacturing system with autonomous entities having the needed level intelligence to be changeable to organize themselves to altered situations, and to identify what new entities will be required. At the same time, a manufacturing system is required to be competitive in order to survive in the markets as well as sustainable to reduce or eliminate unwanted activities and outputs.

2.1 Competitive and sustainable manufacturing

The well-known definition of sustainability is: "The Sustainable Development is development that meets the needs of the present without compromising the ability of future generations to meet their own needs" (World Commission on Environment and Development [WCED], 1987), thereafter (WCED, 1987). This political statement is the root cause for today's key global challenges and related problems that call for a drastic change of paradigm from economic to sustainable development. Competitive Sustainable Manufacturing (CSM) is seen as a fundamental enabler of such change (Jovane, 2009).

Sustainable development has been recently increasingly emphasized around the world; in Europe (Factories of Future Strategic Roadmap and the Manufuture initiative), the USA (Lean and Mean), and Japan (Monozukuri and New JIT). The CSM paradigm widens the classical view of sustainability to interact with the Social, Technological, Economical, Environmental, and Political (STEEP) context (AdHoc, 2009). Sustainable manufacturing is a multi-level approach where product development, manufacturing systems and processes as well as enterprise and supply chain levels need to be considered, with metrics identified for each level (Jawahir et al., 2009).

The CSM is one of the strategic research areas within the Department of Production Engineering (TTE) at Tampere University of Technology (TUT). Figure 1 presents the main areas of the CSM approach, consisting of three main pillars, Sustainable, Lean and Agile Manufacturing. Lean manufacturing aims to combine the advantages of craft and mass production, while avoiding the drawbacks such as the high costs of craft production and rigidity of mass production systems (Womack et al., 1990). For example, the Lean Enterprise Institute (2008) defines Lean manufacturing as "a business system for organizing and managing product development, operations, suppliers, and customer relations that requires less human effort, less space, less capital, and less time to make products with fewer defects to precise customer desires, compared with the previous system of mass production."

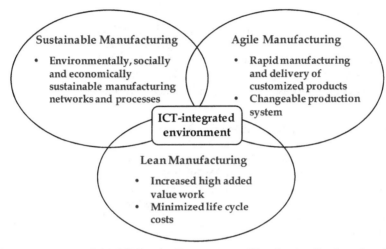

Fig. 1. The cornerstones of the CSM at the Department of Production Engineering (Nylund et al., 2010)

Agile manufacturing can be defined as an enterprise level manufacturing strategy of introducing new products into rapidly changing markets (Nagel & Dove, 1991) and an organizational ability to thrive in a competitive environment characterized by continuous and sometimes unforeseen change (Kidd, 1994). Agile manufacturing highlights the need to adapt to changes in the business environment, and generally agility is defined as ability to react to and take advantage of changes and opportunities, see for example (Sharifi & Zhang, 1999; Gould, 1997).

Sustainable development is the development that meets the needs of the present without compromising the ability of future generations to meet their own needs (WCED, 1987). It consists of three structural pillars namely society, environment, and economy, whilst at the same time it also involves operational aspects such as the consumption of resources, natural environment, economic performance, workers, products, social justice and community development (Jayachandran et al., 2006). When these three pillars of Lean, Agile, and Sustainable are considered as one system, Lean emphasized the stability of a system that can be referred as the autonomy while agility adds the needed capability to change to new situations, therefore focusing more on the collaboration. These two have their main focus on economic issues while sustainability adds the viewpoints of energy and environmentally friendly manufacturing.

2.1.1 Changeability in manufacturing systems

Wiendahl et al. (2007) suggest changeability as an important factor in the competitiveness of manufacturing companies in addition to the classical factors of cost, quality, and time. Changeability is defined on the five structuring levels of an enterprise: changeover ability, reconfigurability, flexibility, transformability and agility. Agility, which was discussed in the context of CSM, is seen from a manufacturing enterprise level and refers to the ability of an enterprise to effect changes in its systems, structure and organization (Gunasekaran & Yusuf, 2002).

Transformability is changeability at a factory level. It includes, for example, facilities, organization and employees. The whole factory is oriented towards the market to offer the right products and services (Wiendahl et al., 2007). Into a detailed level of manufacturing activities the term changeover ability is used. It is related to single workstations that perform manufacturing processes in order to manufacture product features.

Reconfigurability and flexibility are the most widely examined structuring levels of changeability in the context of manufacturing systems. An FMS is configured to deal with part variations within its scope. The functionality and capacity of FMSs are pre-designed, while flexibility is inherent and built-in a priori (ElMaraghy, 2005). Because of the fixed flexibility of FMS, it is not flexible enough for rapid and cost-effective reconfiguration in response to changing markets (Mehrabi et al., 2000). An RMS is composed of general-purpose hardware and software modules that are reused in reconfiguration tasks. Modules are replaced or added only if necessary. An RMS has the ability to change capacity and functionality to bring about the needed flexibility, i.e. to bring about exactly the functionality and capacity needed exactly when needed (Koren, 1999).

2.2 Support from digital manufacturing

The tools and principles of digital manufacturing, factories, and enterprises can offer significant value to all aspects of manufacturing systems during their life cycles. However, there are no commonly used or agreed definitions for those, but they usually share the idea of managing the typically isolated and separate manufacturing activities as a whole by the means of Information and Communications Technology (ICT) (Nylund and Andersson, 2011). Typical examples often found from the definitions, based on literature, are (see, for example: Bracht & Masurat, 2005; Maropoulos, 2003; Souza et al., 2006):

* An integrated approach to develop and improve product and production engineering technologies.
* Computer-aided tools for planning and analysing real manufacturing systems and processes.
* A collection of new technologies, systems, and methods.

Typical tools and principles of digital manufacturing on different structuring levels are, for example (Kühn, 2006):

* Computer-aided technologies, such as computer-aided design (CAD) and computer-aided manufacturing (CAM), e.g. offline programming for virtual tool path generation to detect collisions, analyse material removal and optimise cycle times.
* Visual interaction applications, e.g. virtual environments and 3D-motion simulations that offer realistic 3D graphics and animations to demonstrate different activities.
* Simulation for the reachability and sequences of operations as well as internal work cell layout and material handling design. These include, for example, realistic robotics simulation (RRS) and ergonomics simulation.

- Discrete event simulation (DES) solutions including the need for and the quantity of equipment and personnel as well as evaluation of operational procedures and performance. DES can also be focused on e.g. factories and supply chain or network sales and delivery processes as well as to complex networked manufacturing activities, including logistical accuracy and delivery reliability of increasing product variety.

The above are examples of typical application areas of digital manufacturing. In each case, the activities rely on up-to-date and accurate information and knowledge. The total information and knowledge of a manufacturing system can be explained with explicit and tacit components (Nonaka and Takeuchi, 1995). The explicit part of the knowledge can be described precisely and presented formally in ICT-systems. The skills of humans are explained as the tacit dimension of knowledge, which, presented digitally, may lead to unclear situations and can be wrongly understood. The importance of the transformation from tacit to explicit knowledge has been recognized as one of the key priorities of knowledge presentation (Chryssolouris et al., 2008).

Challenges exist both in the autonomous and collaborative parts of the digitally presented manufacturing entities. The internal part should include only the needed information and knowledge to fully describe the autonomous activities while the collaboration mostly relies on effective sharing of information and knowledge and therefore both the communication language and content should be described formally. Effective knowledge management consists of four essential processes: creation, storage and retrieval, transfer, as well as application, which are dynamic and continuous phenomenon (Alavi and Leidner, 2001). Examples of the application areas of the digital part are:

- Email messages, Internet Relay Chat (IRC), Instant Messaging, message boards and discussion forums.
- More permanent information and knowledge derived from the informal discussions, stored in applications such as Wikipedia.
- Internet search engines and digital, such as dictionaries, databases, as well as electronic books and articles
- Office documents, such as reports, presentations, as well as spreadsheets and database solutions.
- Formally presented information systems, such as Enterprise Resource Planning (ERP), Product Data Management (PDM), and Product Lifecycle Management (PLM).

The importance of the possibilities offered by ICT tools and principles is ever more acknowledged, not only in academia, but also in industry. The Strategic Multi-annual Roadmap, prepared by the Ad-Hoc Industrial Advisory Group for the Factories of the Future Public-Private Partnership (AIAG FoF PPP), lists ICT as one of the key enablers for improving manufacturing systems (AdHoc, 2010). The report describes the role of ICT at three levels; smart, virtual, and digital factories.

- Smart factories involve process automation control, planning, simulation and optimisation technologies, robotics, and tools for competitive and sustainable manufacturing.
- Virtual factories focus on the value creation from global networked operations involving global supply chain management.

- Digital factories aim at a better understanding and the design of manufacturing systems for better product life cycle management involving simulation, modelling and management of knowledge.

Both digitally presented information and knowledge as well as computer tools and principles for modelling, simulation, and analysis offer efficient ways to achieve solutions for design and development activities. General benefits include, for example:

- Experiments in a digital manufacturing system, on a computer model, do not disturb the real manufacturing system, as new policies, operating procedures, methods etc. can be experimented with and evaluated in advance in a virtual environment.
- Solution alternatives and operational rules can be compared within the system constraints. Possible problems can be identified and diagnosed before actions are taken in the real system.
- Modelling and simulation tools offer real-looking 3D models, animations, and visualisations that can be used to demonstrate ideas and plans as well as to train company personnel.
- Being involved in the process of constructing the digital manufacturing system tasks increases individuals' knowledge and understanding of the system. The experts in a manufacturing enterprise acquire a wider outlook compared to their special domain of knowledge as they need to gather information also outside their daily operations and responsibilities.

3. Structure of manufacturing entities and systems

The proposed structure of manufacturing systems consists of manufacturing entities as well as their related domains and activities. An entity, being autonomous, is something that has a distinct existence and can be differentiated from other entities. The term 'entity' has similarities to other terms, such as: object, module, agent, actor, and unit. A domain is an expert area in which two or more entities are collaborating. Domains have certain roles in the system and their own responsibilities and specific objectives. An activity is a set of actions that accomplish a task that is related to the entities and domains, as well as to their context.

3.1 Structure of manufacturing entities

Figure 2 illustrates the general viewpoints of the proposed structure of manufacturing entities. The structure is explained with internal structure of individual manufacturing entities. It is derived from the principles behind the term 'holon' and the concept of Holonic Manufacturing Systems (HMS). The term holon comes from the Greek word 'holos', which is a whole and the suffix '–on', meaning a part. Therefore the term holon means something that is at the same time a whole and a part of some greater whole (Koestler, 1989).

In HMS, holons are autonomous and co-operative building blocks of a manufacturing system, consisting of information processing part and often a physical processing part (Van Brussel et al., 1998). In this approach, the information part is divided into digital and virtual parts differentiating the digitally presented information and knowledge from the computer models representing the existing or future possible real manufacturing entities. The digital part barely exists as clearly consisting separate part. It can be distributed in several information systems both globally and locally and in information rich computer models, the virtual parts of the manufacturing entities.

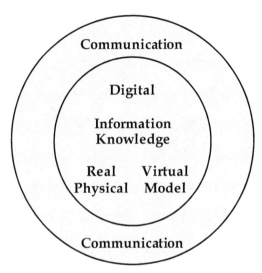

Fig. 2. Internal structure of manufacturing entities

The digital, virtual, and real parts combined present the autonomy of a manufacturing entity. The communication part is responsible of both the language and content of the messages between manufacturing entities. Therefore, it enables the manufacturing entities to collaborate with each other (Nylund & Andersson, 2011). As the autonomous entities exist distributed, independently from each other, they can be developed separately. At the same time, the communication part enables the investigation of the entities in an integrated fashion, and to develop the whole system they form.

The division into digital, virtual, and real is intentionally missing the tacit dimension, as it is intended to be used in decision making processes by humans, based on their skills and knowledge. At the end, the humans are the ones that are making the decisions, or are the ones that are creating the decision making mechanisms.

3.2 Structure of manufacturing systems

A manufacturing system consists of manufacturing entities with different roles as well as their related domains and activities. Figure 3 shows a general presentation of manufacturing entities of products, orders, and resources as well as their connecting domains of process, production, and business. The focus is on the manufacturing activities that are related to the transformation of raw material to finished products and their associated services as well as the flow of information and knowledge that is related to the physical manufacturing of customer orders.

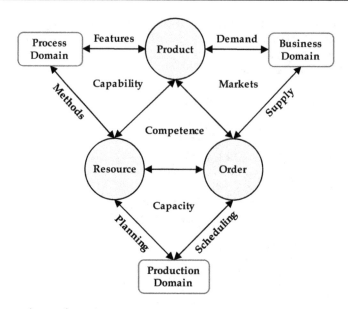

Fig. 3. Structure of manufacturing systems

The proposed structure is loosely based on the HMS reference architecture Product-Resource-Order-Staff Architecture (PROSA) (Van Brussel et al., 1998). The PROSA explains the relations between the entities with the information and knowledge they exchange while in this approach the relations are explained with activities occurring between the entities. Brief descriptions of the entities and domains are:

- *Products* represent what the manufacturing system offers to its customers. The characteristics of the products specify the requirements for the manufacturing system, i.e. what the system should be able to do.
- *Resources* embody what is available to manufacture the products. The characteristics of the resources determine what kinds of products can be manufactured.
- *Orders* represent instances of products that are ordered by customers. They define the volume and variation requirements of the products ordered, as well as the capacity and scalability requirements for the manufacturing system.
- *The process domain* represents the capabilities that are needed to manufacture the products. It connects the development activities of products and resources.
- *The production domain* defines the capacity and scalability to manufacture changing volumes and variations in customer orders. It handles the material and information flow of the manufacturing system.
- *The business domain* is responsible for markets, i.e. for the right products being available for the customers to gain enough orders.

3.3 Structuring levels in manufacturing

A fractal is an independently acting manufacturing entity that can be precisely described (Warnecke, 1993). Fractals are structured bottom-up, building fractals of a higher order.

Entities at the higher levels always assume only those responsibilities in the processes which cannot be fulfilled in lower order (Strauss & Hummel, 1995). This is similar to holons and holarchies, as at every fractal level of holons the level above is the holarchy of the holons at a lower level. Similarly, the autonomy of the holons is not considered in the holarchy, but instead dealing with and organizing the co-operation of the holons is the responsibility of the holarchy.

In Figure 4, four different structuring levels, manufacturing units, stages, plants, and networks, are distinguished.Manufacturing units correspond to individual machine tools that have certain manufacturing methods. The units are designated to manufacture the features of work pieces that have similarities in, for example, size and shape as well as tolerances and material properties. Typical areas are computer-aided design (CAD) and computer-aided manufacturing (CAM), e.g. offline programming for virtual tool path generation to detect collisions, analyse material removal and optimise cycle times (Kühn, 2006).

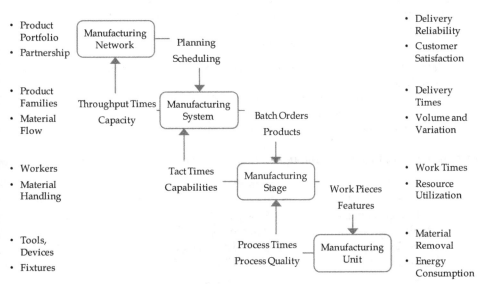

Fig. 4. Examples of structuring levels of manufacturing and their connections

Manufacturing stages are physical or logical manufacturing areas, e.g. manufacturing cells, consisting of one or more manufacturing units and their co-operation. Additionally, the manufacturing stages include internal material handling in moving the work pieces between the manufacturing units as well as buffers and stocks to hold batches of the work pieces. In manufacturing stages the focus can be on simulation for the reachability and sequences of operations as well as internal work cell layout and material handling (Kühn, 2006).

Manufacturing plants are composed of manufacturing stages, warehouses for storing the products as well as internal logistics to transfer material between the stages and material storing areas. They typically correspond to factories and have customers who can be other companies or internal customers, such as an assembly plant. Typical simulation issues concern the layout design and material flow analysis as well as planning and controlling the manufacturing activities. Simulation studies on a manufacturing plant level are usually conducted using discrete event simulation (DES) including the need for and the quantity of equipment and personnel as well as evaluation of operational procedures and performance. Manufacturing networks consist of factory units, which can exist globally. One of the key differences between plants and networks is that entities in the network often belong to different companies that may have contradictory goals in their strategies. Simulation can be focused on traditional supply chain sales and delivery processes as well as to complex networked manufacturing activities, including logistical accuracy and delivery reliability of increasing product variety.

4. Digital manufacturing support for manufacturing activities

A digitally presented manufacturing system contains the information and knowledge of manufacturing entities and activities that it is reasonable to represent in a digital form. This, at its best, makes possible efficient collaboration between all the manufacturing activities and related parties. The discussion on digital manufacturing support is based on a previously developed framework for extended digital manufacturing systems (EDMS). An EDMS can briefly be defined as follows (Nylund and Andersson, 2011):

• an integrated and collaborative environment for humans, machines, and information systems to act and interact;
• to enhance the research, development and management activities of products, production systems, and business processes,
• supporting knowledge-intensive decision-making in the entirety of their lifecycles.

4.1 From ideas to innovative solutions

Figure 5 represents a process from ideas and the need for change to innovative solutions. It consists of a chain of activities where the results evolve towards more precise solutions. Each phase has its enablers as inputs and the activity creates results as outputs. The results affect the enablers in the following phases of the process. The process is also iterative as it is possible to go back to previous phases in order to change or refine them. The need for change can arise, for example, from social, technological, economic, environmental, and political aspects.

The changes can also derive from voluntary ideas that are seen to improve the competence of the system. If the process has not been developed previously, the current system has to be analysed to create the digital information and knowledge of what currently exists. The synthesis of the existing system and possible changes form the new requirements for the future system. The combination of feasible new possibilities and existing capabilities forms the solution principles. The results are digital entities and abstract and conceptual descriptions, including the objectives and preliminary properties of the future system.

When the descriptions evolve towards a more detailed level, possible technologies can be investigated, resulting in alternative solutions. The solution alternatives can be modelled as virtual entities that include, in addition to their digital description, for example, 3D models with their own operating rules, motion, and behaviour. Combining the existing and new virtual entities forms a rough simulation model. The solution that is implemented has to be verified to make sure that the behaviour and co-operation of the entities in the system are modelled correctly. The verified simulation model can be used to run test experiments. By analysing the results from the simulation model and comparing them with known or predicted outcomes, the behaviour of the simulation model can be validated. When the simulation model is verified and validated, it can be used for manufacturing experiments. The experiments are used to analyse the behaviour of the system, and can lead towards innovative solutions.

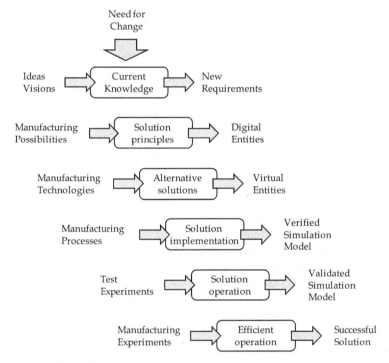

Fig. 5. The process from ideas to innovative solutions

4.2 Manufacturing process and flow development

Figure 6 shows a theoretical example of process and flow development. The manufacturing process part corresponds to the process domain, presented in Figure 3, where the capabilities for a manufacturing network are developed. The part of the manufacturing flow presents the production domain in Figure 3, aiming for the right capacity and scalability of the manufacturing network to meet the customer demands. The existing capabilities are combined with new possibilities, requirements, and constraints in the production network creating the synthesis of existing and what new capabilities will be required. These derive from, for example, new possible markets, customers, and competition i.e. what is important in the future that the current capabilities cannot fulfil. The new possible capabilities are tested virtually using computer-aided technologies in connection with the digitally presented information and knowledge.

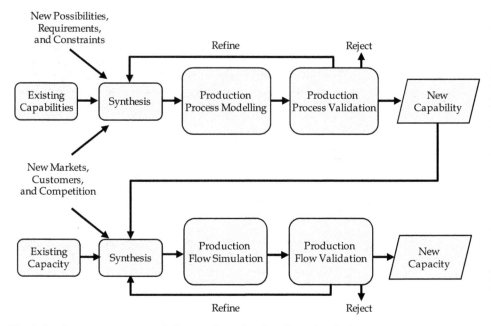

Fig. 6. Production process modelling and production flow simulation.

The resulted new capability is validated both to ensure that it does what it is supposed to do and that it meets the performance requirements, such as cost, quality, and time efficiency as well as the social and environmental issues. The production flow simulation in Figure 6 follows the same idea as the production process modelling. The new capability can add to the total capabilities of the network if something new is implemented, or change the existing capabilities if something already existing is reconfigured. It is not enough that all the needed capabilities exist.

The production flow simulation aims to define how much capabilities are required to produce the changing volume and variation of customer orders at the right time. Typical areas are the controlling, planning and scheduling of the activities. To investigate the production process modelling in more detail, five categories between product requirements and resource capabilities can be recognized, see Figure 7:

- Existing capability: The capabilities exist for all of the product requirements without any need for changes to the system. The products can be manufactured as the service requests have service providers.
- Possible existing capability: At least some of the product requirements need further investigation as to whether the capabilities exist or not. The requirements are close to the existing capabilities and, using modelling and simulation, the capabilities can be verified.
- Capability after reconfiguration: There is no existing capability but it may be possible to reconfigure the system so that it has the capabilities. By modelling the reconfigured system the possibility can be verified.

- Capability after implementation: The system does not have the needed capability. It may be possible if new capabilities are added to the system. Again this can be verified using modelling and simulation.
- No capability: The result may also be that there are no capabilities and they cannot be implemented either. This leads to the need for an alternative solution, which leads to a result that fits into one of the first four categories.

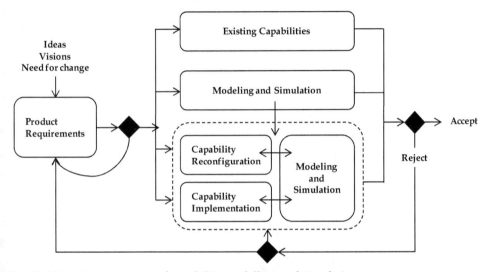

Fig. 7. Alternative outcomes of capability modelling and simulation

When it is known that the capabilities exist for all the product requirements, the efficiency of the capabilities still needs to be evaluated against factors such as cost, quality, and time. It has to be decided if the solution alternative is good enough. It can be further investigated in the capacity loop or it can be rejected and sent back to the capability loop. If all the needed capabilities exist, the capacity of the system has to be checked. The same five categories can be used in capacity evaluation. If it is known that there is enough capacity, nothing else has to be done. Modelling and simulation can be used to verify that there is enough capacity. It can also be used in capacity reconfiguration and implementation issues. Modelling and simulation of capacity has the same constraints as in the case of capabilities. The capacity for existing volume and variation still has to exist when new products are considered as an addition to existing products. In the capacity loop, the solution can be accepted or rejected, as in the capability loop. If the solution is rejected, it can be sent back to the capability loop or further back into the design requirements loop.

4.3 Manufacturing system operation

Operation of a manufacturing system can be viewed from the time dimensions of past, present, and future. The past represents what has happened i.e. it can be said to be the digital memory of the system. The time dimension of the present, what is happening now, is used to operate the current system by monitoring the state of the system and comparing it to the desired state. The future dimension makes it possible to plan future manufacturing activities ahead and to compare different changes in strategies. Figure 8 shows the connection of the time dimensions into the operation of manufacturing systems.

The past presents the data collected from the system activities when they happened. It can be used to analyse previous manufacturing activities in order to find out what happened and the reasons why it happened. In finding the root causes for phenomena, the system can learn from its past and prevent unwanted situations in the future. Rules for the autonomy of the manufacturing entities, as well as for their collaboration, can be enhanced and new rules can be created. The present here means the near future, where no major changes are planned.

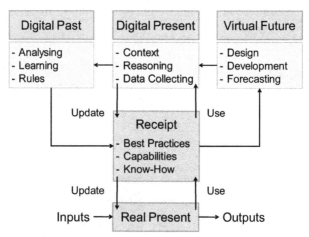

Fig. 8. Digitally co-existing past, present, and future time dimensions

It is, for example, the use of existing resources and the planning and scheduling of customer orders that have already been placed. In the present the digital and real existences co-exist. As the system operates the activities are logged, creating new history data to be analysed and to aid decision-making. The state of the real manufacturing system can be seen in the digital manufacturing system and actions can be taken with the state of the system as a starting point.

The dimension of the future relies on the information and knowledge gathered from the system previously. Future design and development decisions are syntheses of existing capabilities and requirements combined with future goals and possibilities. The viewpoint of the future can be divided into tactical decisions and visions. Tactical decisions consider the near future into which the manufacturing system is heading. Future visions are similar to tactical decisions, the difference being the time horizon.

The outcome of future visions is more obscure but there are more possibilities to be investigated. The information and knowledge from analysing the past, collecting data from the present, and forecasting the future is stored in the form of receipts. A receipt holds the capabilities of a system, constantly updating and refining the best practices in conjunction with human skills and know-how. The receipts are the basis of the operations in the real present, the only time dimension in the real world.

4.4 Continuous analysis and improvement

A manufacturing system can be seen as multiple autonomous manufacturing entities interacting and co-operating in a complex network of manufacturing activities. The activities are explained as services, which hold the information and knowledge needed to explain the manufacturing activities. It is required that the activities are known exactly, in that they are understood by all related parties.

Describing the activities as services in a digital format creates a formal way to present the services. This makes possible efficient collaboration in a digital manufacturing system between entities that can be humans, machines, or information systems. The information and knowledge is kept as the autonomous property of the manufacturing entities and the communication between the entities includes only the information that is needed to fully describe the collaboration activity.

The communication between the manufacturing entities is loosely based on service-oriented architecture (SOA), which consists of self-describing components that support the composition of distributed applications (Papazoglou & Georgakopoulos, 2003) enabling the autonomous manufacturing entities to negotiate and share their information and knowledge. The basic conceptual model of the SOA architecture consists of service providers, service requesters, and service brokers (Gottschalk, 2000). The roles of manufacturing entities in a digital manufacturing system based on SOA are briefly explained as follows:

- Service requesters are typically product entities when they are realized as order entities. The order entities call on the services they require to be manufactured.
- Service providers include the manufacturing resource entities which have the capabilities needed to provide the services that are requested.
- Service broker plays a role of an actor that contains the rules and logics of using the services. Its function is to find service providers for the requesters on the basis of criteria such as cost, quality, and time.

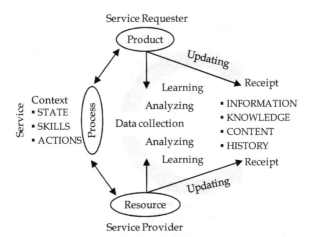

Fig. 9. An example of a service between a service provider and a service requester

Figure 9 shows an example of a service occurring in the process domain between products and resources. A service consists of two different entities, i.e. the product and resource entities having the roles of service requesters and providers. The actual service, being the manufacturing activity, is twofold, consisting of a context and receipt. The context is the environment, real or virtual, where the service takes place, whilst the receipt is the digital description of the service. The product entity requests a service, which is provided by the resource entity. The service, whether it is happening in a virtual or real environment, has a certain context that is in a certain state. The state is a basis for the actions happening during the service, and the result is based on the skills of the service provider. During the service data are collected from the process. The collected data are analyzed, forming information that is the basis for learning from the service. When something is learned, it is used to update the receipt, which will be the basis for future services.

When a certain product entity uses a service provided by a certain resource entity, the data collection, analysis, learning, and updating phases include adding the same data and information to the knowledge of both entities. The knowledge of a resource entity is updated with several product entities using the services it provides. In a similar fashion, the knowledge of a product entity consists of all the services it requests. A service can be seen as a hierarchy in which a service on the upmost level divides iteratively into multiple sub-services until the level on which the individual part features are requested. This means that an entity requesting a service gets information about the possible service provider entities, but it does not know how the service request is fulfilled. For example, a service request for the manufacturing of a product is a request on the macro level. The macro level service request is divided into multiple sub-services on the meso level and the meso-level service makes similar requests on the micro level. The upper level only needs the information about whether the service request can be fulfilled or not. The hierarchy of the services may be limited by the service requester as it may state special requirements for the service that limit the selection of possible providers. For example, a customer may require certain parts of ordered products to be manufactured in a specific manufacturing plant.

5. Academic research environment

Several of the theoretical issues discussed in this chapter have been implemented into an academic research environment of which real machinery exists in the TTE heavy laboratory, see Figure 10. The digital part of the environment has been constructed as a modular ICT architecture and the virtual part exists as simulation and calculation models. The aim of the environment is to offer a research platform that can be utilised in:

- Designing, developing and testing current and future research topics.
- Prototyping possible solutions for industrial partners in ongoing research projects.
- Utilizing it as an educational environment for university students and company personnel to introduce the latest results in the area of intelligent manufacturing.

The initial version of the environment was introduced during the Tampere Manufacturing Summit seminar, which was held in Tampere, Finland, in June 2009. Since then the environment has been discussed in scientific research papers as well as in seminar and conferences.

Fig. 10. The research environment in TTE heavy laboratory

5.1 General description of the research environment

The research environment consists of typical manufacturing resources and work pieces as physical entities. The resources of the research environment, offering different manufacturing capabilities are, see Figure 11:

- Machine tools (a lathe and a machining centre) for machining operations.
- Robots for material handling and robotized machining operations.
- Laser devices for e.g. machining and surface treatment.
- A punch press, existing only virtually, for the punching of sheet metal parts. The real punch press is located at a factory of an industrial project partner company.

Fig. 11. The machine tools and devices of the research environment

The work pieces, which can be manufactured in the environment, are fairly simple cubical, cylindrical, and flat parts in shape. They have several parameterized features that can be varied within certain limits, e.g. dimensions (width, length, and depth), number of holes, internal corner radiuses, and sheet thickness. The main reasons for the parameterization are, firstly, that the number of different parts can be increased with the variation without having a large number of different types of parts. Secondly, the parameters can be set in a way where changing the parameters also requires capabilities of different kind i.e. different manufacturing resources are required. This gives more opportunities to compare alternative ways to manufacture the work pieces based on selected criteria, such as the cheapest or fastest way to manufacture a work piece.

5.2 Viewpoints of the environment

The research environment can be seen from the digital, virtual, and real viewpoints. Figure 12 shows the digital, virtual, and real views of the whole research environment. The environment can be viewed from three different structuring levels; the whole environment, machining and robot cells, as well as the individual machine tools and robots. The real part of the environment exists in a heavy laboratory and is divided into two main areas, one including the robots and laser devices, and the second consisting of the machine tools. The real manufacturing entities on each structuring level have their corresponding computer models and simulation environments as their virtual parts.

The information and knowledge of the environment is stored in local databases of the manufacturing entities as well as in a common Knowledge Base (KB) for the whole environment, those presenting the digital part of the environment. The actual connection is enabled by and executed via the KB, see (Lanz et al., 2008), as all communication activities use or update it. The KB is the base for the ICT-related research and development activities of the research environment. It is a system where the data of the environment can be stored and retrieved for and by different applications existing in the environment.

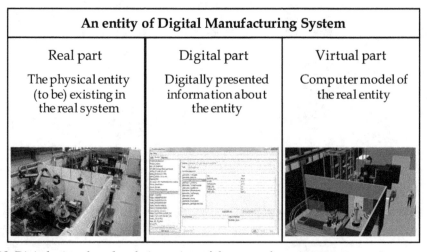

An entity of Digital Manufacturing System		
Real part	Digital part	Virtual part
The physical entity (to be) existing in the real system	Digitally presented information about the entity	Computer model of the real entity

Fig. 12. Digital, virtual, and real viewpoints of the research environment

5.3 Scenarios for manufacturing tests

Figure 13 presents an overall view of the process of digital, virtual and real manufacturing tests that can be performed using the research environment. The product and manufacturing information and knowledge holds what is known about the manufacturing resources of the environment and products that have been manufactured in the environment. The manufacturing methods of the resources are described as capabilities of the research environment i.e. what is known that can be manufactured within the environment. The product requirements are described similarly including all manufacturing features of products that have been previously manufactured. When the ability to manufacture a new product will be examined, firstly a CAD model of the product is required. The CAD model will be analyzed using a feature recognition property of the research environment. For each product feature, a service request is created. The request is sent to the process planning part of the environment to compare the requirements of the new request to existing capabilities of the environment. If a suitable service exists i.e. there exists a process plan for the product feature, the result will be an existing service and no further examination is required. Otherwise, the new service request will be tested for its manufacturability.

The manufacturing tests can roughly be divided into three categories being digital, virtual and real test manufacturing. The digital test manufacturing is basically comparing a set of parameterized values of the service request to the formally described capabilities of the manufacturing resources. The process is quite rapid as it is happening in a computer and no visualization or animation is required. It is the most favourable choice if time is limited.

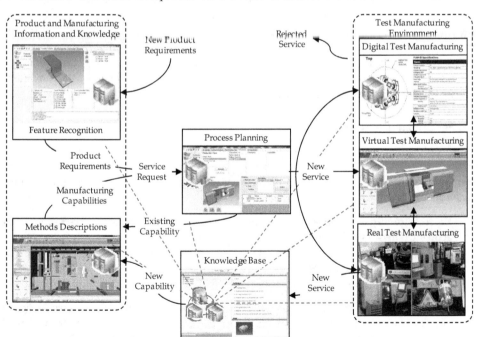

Fig. 13. An overall view of the manufacturing tests of the research environment.

The second choice would be a virtual test manufacturing i.e. typically modelling and simulation. It requires more time as human interaction is required during the process. The time required is dependent if existing simulation models can be used or new simulation models need to be constructed. In the case, where the existing simulation models cannot be used, new ones are required to be built. The creation of a new simulation model may be to reconfigure the existing virtual system to meet the new requirements, or implementing something new into the system if the system does not have all the required capabilities or manufacturing resources. In these alternatives, the test manufacturing is still carried out with computers i.e. it does not interrupt the use of the real manufacturing resources.

The real test manufacturing will be the choice if the digital or virtual manufacturing tests are not accurate enough to fully trust or understand the results gathered from the test. The real test manufacturing requires the physical resources and the time used will reduce the time for daily operations to manufacture customer orders. In some cases it is also reasonable to conduct additional tests with real manufacturing resources to reduce the risk of implementing fault processes. The responsibility of selecting, whether digital or virtual test manufacturing would be enough, is to be determined by humans, based on their skills and knowledge of the matter in hand, and has to be evaluated separately for each time a decision needs to be made. After the manufacturing tests have been conducted, the alternative is either a rejected or accepted new service. The result of rejected service could happen if the product feature cannot be manufactured within the system, or even if it could be manufactured, it is e.g. too expensive, uses too much time or does not output desired quality. In these cases, the results can be fed back to the product development to consider it the feature can be redesigned. In the case where the new service is accepted, it is added as a new capability of the environment and new process plan will be created. This will increase the known capabilities of the environment as each test manufacturing test adds new information and knowledge to the digital part of the environment, which will be available for the future test manufacturing cases.

5.4 Performance metrics

The measurements of the manufacturing environment can be divided into direct and indirect measures. The direct measurements are achieved using the sensors and measurement devices in the environment, and the metrics can be calculated immediately. Examples of the direct measurements are:

- Process quality assurance, a real time measurement using force, acceleration, and acoustic emission (AE) sensors.
- Process stability monitoring following the electricity variation of the robot servomotor caused by the cutting forces.
- Energy consumption monitoring using a Carlo Gavazzi EM21 72D energy meter.

In the case of the indirect measurements the logged data are stored in the history section of the KR. The data can be analyzed and to create the desired performance metrics. Table 1 summarizes the performance metrics from the viewpoints of manufacturing operation, production supervising, and business management.

Performance metric	Manufacturing Operator	Production Supervisor	Business Management
Cost	Continuous improvement to reduce the cost per part	Using the most cost-efficient production choices	The gain more profit and offer cheaper products to customers
Quality	To assure the manufacturing process efficiency and stability	Delivery reliability and Just-In-Time manufacturing	Improved customer satisfaction and decreased reclaims
Material consumption	To use near-net shape blank material	To reduce waste, material and energy use to meet the sustainability requirements	Meeting the requirements of legislation and expectations of the society by reducing the unwanted effects
Waste			
Energy consumption	To have real energy consumption results		
Production load and time metrics	To reduce the time per part and to update any changes in the manufacturing process times	To efficiently plan and schedule production to utilize the capacity of the system	To know how much customer orders can be placed and to give more precise delivery dates
Resource utilization			

Table 1. Different views to utilize the performance metrics

6. Conclusion

This Chapter discussed on the possibilities of digital manufacturing to support efficient activities of designing, developing and operating manufacturing systems. A structure of individual manufacturing entities and whole systems was proposed. Describing entities of a manufacturing system as independent, yet closely related existences of digital, virtual and real enables more efficient and effective manufacturing activities from early conceptual ideas to successful solutions. Even when describing the manufacturing entities independently, they are required to be closely integrated with each other and that can be done via domains of manufacturing related activities of products, resources, and business. Again, when the entities and domains are combined, the integrated fashion should also be invested separately in different structuring levels of manufacturing, yet again closely integrated between the structuring levels

By keeping the entities the same during their whole lifecycle reduces the loss of information and knowledge and enables more efficient manufacturing activities. These we discussed from several aspects i.e. a path from early ideas and needs to efficient solutions, development of manufacturing processes and flow, as well as how a system can learn from its daily operations by collecting and analysing data from the activities that can help in learning thus improving the way to do things in future.

An academic research environment was discussed on how these theoretical aspects can be implemented into a manufacturing environment. As the environment is constantly developed, some of the issues have been fully implemented while some other areas remain as a future of the environment. This is due to the fact that the current and future research topics lead the development of the environment.

7. Acknowledgment

The research presented in this paper is co-financed by Tekes (the Finnish Funding Agency for Technology and Innovation), TUT Foundation and several major companies in Finland. The authors would like to thank all the colleagues involving on the development of the academic research environment.

8. References

Ad-hoc Industrial Advisory Group. (2010). Factories of the Future PPP - Strategic Multi-annual Roadmap.

Alavi, M. & Leidner, D.E. (2001). Knowledge Management and Knowledge Management Systems: Conceptual Foundations and Research issues. *MIS Quarterly*, Vol. 25, pp. 107-136, ISSN: 0276-7783

Andersson, P.H. (2007). FMS in 2010 and beyond. *Tampere Manufacturing Summit 2007*, Tampere, Finland, June 2007

Bracht, U. & Masurat, T. (2005). The Digital Factory between vision and reality. *Computers in Industry*, Vol. 56, pp. 325-333, ISSN: 0166-3615

Chryssolouris, G., Mavrikios, D., Papakostas, N., Mourtzis, D., Michalos, G. & Georgoulias K. (2008). Digital manufacturing: history, perspectives, and outlook. *Proceedings of the Institution of Mechanical Engineers Part B: Journal of Engineering Manufacture*, Vol. 222, No. 5, pp. 451-462, ISSN (printed): 0954-4054. ISSN (electronic): 2041-2975

ElMaraghy, H.A. (2005): Flexible and reconfigurable manufacturing systems paradigms. *International Journal of Flexible Manufacturing Systems, Special Issue on Reconfigurable Manufacturing Systems*, Vol. 17, pp. 261-276, ISSN (printed): 0920-6299. ISSN (electronic): 1572-9370

Gottschalk, K. (2000). Web Services architecture overview. IBM developerWorks, Whitepaper

Gould, P. (1997). What is agility?. *Manufacturing Engineer*, Vol. 76, No. 1, pp. 28-31, ISSN

Gunasekaran, A. & Yusuf, Y.Y. (2002). Agile manufacturing: a taxonomy of strategic and technological imperatives. *International Journal of Production Research*, Vol. 40, No. 6, pp. 1357-1385, ISSN (printed): 0020-7543. ISSN (electronic): 1366-588X

Jawahir, I.S., Badurdeen, F., Goldsby, T., Iyengar, D., Gupta, A., Metta, H., Stovall, C. & Ladd. C. (2009). Assessment of Product and Process Sustainability: Towards Developing Metrics for Sustainable Manufacturing. *NIST Workshop on Sustainable Manufacturing*, Gaithersburg, MD, USA, October 2009

Jayachandran, R., Singh, R., Goodyer, J. & Popplewell, K. (2006). The design of a sustainable manufacturing system: A case study of its importance to product variety manufacturing. *Intelligent production machines and systems, 2nd I*PROMS virtual international conference*, ISBN 13: 978-0-08-045157-2, Pham, D.T., Eldukhri, E.E. & Soroka A.J. (eds.), pp.650-656

Jovane, F. (2009). The Manufuture Road towards Competitive and Sustainable HAV Manufacturing, *Tampere Manufacturing Summit*, Tampere, Finland, June 2009

Kidd, T. (1994). *Agile Manufacturing: Forging new frontiers*. Addison-Wesley Reading, 1st edition, ISBN-13: 978-0201631630, Boston, MA.

Koestler, A. (1989). *The Ghost in the Machine*. Arkana Books, ISBN-13: 978-0140191929, London, UK

Koren, Y., Heisel, U., Jovane, F., Moriwaki, T. Pritschow, G., Ulsoy, G. and Van Brussel H. (1999). Reconfigurable manufacturing systems. *CIRP Annals*, Vol. 48, No. 2, pp. 527-540, ISSN: 0007-8506

Kühn, W. (2006). Digital Factory - Integration of simulation enhancing the product and production process towards operative control and optimization. *International Journal of Simulation Modelling*, Vol. 7, No. 7, pp 27-39, ISSN: 1726-4529

Lanz, M., Kallela, T., Velez, G. & Tuokko, R. (2008). Product, Process and System Ontologies and Knowledge Base for Managing Knowledge between Different Clients. *Proceedings of the 2008 IEEE International Conference on Distributed Human-Machine Systems*, pp. 608-513, ISSN: 1094-6977

Lapinleimu, I. (2001). Ideal Factory, Theory of Factory Planning, Produceability and Ideality. Doctoral dissertation, Tampere University of technology, Finland, Publications 328

Lean Enterprise Institute. (2008). *Capsule Summaries of Key Lean Concepts*. accessed September 15, 2011, Available from: http://www.lean.org/WhoWeAre/NewsArticleDocuments/key_lean_definitions.html

Maropoulos, P.G. (2003). Digital enterprise technology-defining perspectives and research priorities. *International Journal of Computer Integrated Manufacturing*, Vol. 16, Nos. 7-8, pp. 467-478, ISSN (printed): 0951-192X. ISSN (electronic): 1362-3052

Mehrabi, M.G., Ulsoy A.G. & Koren, Y. (2000). Reconfigurable manufacturing systems: Key to future manufacturing. *Journal of Intelligent Manufacturing*, Vol. 11, pp. 403–419, ISSN (printed): 0956-5515. ISSN (electronic): 1572-8145

Nagel, P. and Dove, R. (1991). 21st century manufacturing enterprise strategy. Iacocca institute, ISBN-13: 978-0962486630, Bethlehem, PA.

Nonaka, I. & Takeuchi, H. (1995). The Knowledge-Creating Company: How Japanese Companies Create the Dynamics of Innovation. Oxford University Press, ISBN-13: 978-0195092691, Oxford, USA

Nylund, H., Koho, M. & Torvinen, S. (2010). Framework and toolset for developing and realizing competitive and sustainable production systems. *Proceedings of the 20th International Conference on Flexible Automation and Intelligent Manufacturing*, July 2010, Oakland, CA, USA, pp. 294-301.

Nylund, H. & Andersson, P.H. (2011). Framework for extended digital manufacturing systems. *International Journal of Computer Integrated Manufacturing*, Vol. 24, No. 5, pp. 446 – 456, ISSN (printed): 0951-192X. ISSN (electronic): 1362-3052

Papazoglou M.P. & Georgakopoulos, D. (2003). Service Oriented Computing. *Communications of the ACM*, Vol. 46, No. 10, pp .25-28, ISSN: 0001-0782

Sharifi, H. & Zhang, Z. (1999). A methodology for achieving agility in manufacturing organizations: An introduction. *International Journal of Production Economics*, Vol. 62, No. 1-2, pp.7-22, ISSN: 0925-5273

Souza, M.C.F., Sacco, M. & Porto A.J.V. (2006). Virtual manufacturing as a way for the factory of the future. *Journal of Intelligent Manufacturing*, Vol. 17, pp. 725-735, ISSN (printed): 0956-5515. ISSN (electronic): 1572-8145

Strauss, R.E. & Hummel, T. (1995). The new industrial engineering revisited - information technology, business process reengineering, and lean management in the self-organizing "fractal company". *Proceedings of 1995 IEEE Annual International Engineering Management Conference*, Singapore, June 1995, pp. 287-292.

Van Brussel, H., Wyns, J., Valckenaers, P., Bongaerts, L. & Peeters, P. (1998). Reference architecture for holonic manufacturing systems: PROSA. *Computers in Industry*, Vol. 37, pp. 255-274, ISSN: 0166-3615

Warnecke, H.J. (1993). The Fractal Company: A Revolution in Corporate Culture, Springer-Verlag, ISBN-13: 978-3540565376, Berlin, Germany.

WCED (1987) Our Common Future. World Commission on Environment and Development (Brundtland Commission), Oxford University Press, ISBN-13: 978-0192820808, Oxford.

Wiendahl, H.-P., ElMaraghy, H.A., Nyhuis, P., Zäh, M.F. & Wiendahl H.-H. (2007). Changeable Manufacturing - Classification, Design and Operation. *CIRP Annals*, Vol. 56, No. 2, pp.783-809, ISSN: 0007-8506

Womack, J.P., Jones, D.T. & Roos, D. (1990). The machine that changed the world, Rawson, ISBN-13: 978-0060974176

Reliability Evaluation of Manufacturing Systems: Methods and Applications

Alberto Regattieri

DIEM - Department of Industrial and Mechanical Plants, University of Bologna
Italy

1. Introduction

The measurement and optimization of the efficiency level of a manufacturing system, and in general of a complex systems, is a very critical challenge, due to technical difficulties and to the significant impact towards the economic performance.

Production costs, maintenance costs, spare parts management costs force companies to analyse in a systematic and effective manner the performance of their manufacturing systems in term of availability and reliability (Manzini et al. 2004, 2006, 2008).

The reliability analysis of the critical components is the basic way to establish first and to improve after the efficiency of complex systems.

A number of methods (i.e. Direct Method, Rank Method, Product Limit Estimator, Maximum likelihood Estimation, and others (Manzini et. Al., 2009) all with reference to *RAMS* (Reliability, Availability, Maintainability and Safety) analysis, have been developed, and can bring a significant contribution to the performance improvement of both industrial and non-industrial complex systems.

Literature includes a huge number of interesting methods, linked for example to preventive maintenance models; these models can determine the best frequency of maintenance actions, or the optimization of spare parts consumption or the best management of their operating costs (Regattieri et al., 2005, Manzini et al., 2009).

Several studies (Ascher et al..1984, Battini et al., 2009, Louit et al., 2007, Persona et al. 2007) state that often these complex methodologies are applied using false assumptions such as constant failure rates, statistical independence among components, renewal processes and others. This common approach results in poor evaluations of the real reliability performance of components. All subsequent analysis may be compromised by an incorrect initial assessment relating to the failure process. A correct definition of the model describing the failure mode is a very critical issue and requires efforts which are often not sufficiently focused on.

In this chapter the author discusses the model selection failure process, from the fundamental initial data collection phase to the consistent methodologies used to estimate the reliability of components, also considering censored data.

This chapter introduces the basic analytical models and the statistical methods used to analyze the reliability of systems that constitute the basis for evaluation and prediction of the stochastic failure and repair behavior of complex manufacturing systems, assembled using a variety of components. Consequently, the first part of the chapter presents a general framework for components which describes the procedure for the solution of the complete Failure Process Modeling (FPM) problem, from data collection to final failure modeling, that, in particular, develops the fitting analysis in the renewal process and the contribution of censored data throughout the whole process. The chapter discusses the main methods provided in the proposed framework.

Applications, strictly derived from industrial case studies, are presented to show the capability and the usefulness of the framework and methods proposed.

2. Failure Process Modeling (FPM) framework

A robust reliability analysis requires an effective failure process investigation, normally based on non-trivial knowledge about the past performance of components or systems, in particular in terms of failure times. This data collection is a fundamental step. The introduction of a Computer Maintenance Management System - CMMS and of a Maintenance Remote Control System (Persona et al., 2007) can play an important role. Ferrari et al. (2003) demonstrate the risk due to a small data set or due to hasty hypothesis often considered (e.g. constant failure rate, independent identically distributed failure times, etc.).

Literature suggests different frameworks for the investigation of the failure process modeling of components and complex systems, generally focused on a particular feature of problem (e.g. trend tests in failure data, renewal or not renewal approach, etc.).

In this chapter a general framework is proposed considering all the FPM process from data collection to final failure modelling, also considering the contribution of censored data.

Figure 1 presents the proposed framework (Regattieri et al., 2010). Data collection is the first step of the procedure. This is a very important issue, since the robustness of analysis is strictly related to the collected data set. Both failure times and *censored* times are gathered. Times to failure are used for the failure process characterization and censored times finally enrich the data set used for the definition of the parameters of failure models, thus resulting in a more robust modeling.

In general, considering a population of components composed by m units, each specific failure (or inter-failure) time can be found. The result is represented by a set of times called $X_{i,j}$, where i^{th} represents the time of failure of the jth unit: there is a complete data situation in this case, that is, all m unit failure times are available.

Unfortunately, frequently this is not a real situation, because a lot of time and information would be required. The real world test often ends before all units have failed, or several units have finished their work before data monitoring, so their real working times is unknown. These conditions are usually known as *censored data situations*.

Technically, censoring may be further categorized into:

1. Individual censored data
    All units have the same test time t*. A unit has either failed before t* or is still running (generating censored data);
2. Multiple censored data

Test times vary from unit to unit. Clearly, failure times differ but there are also different censoring times. Censored units are removed from the sample at different times, while units go into service at different times.

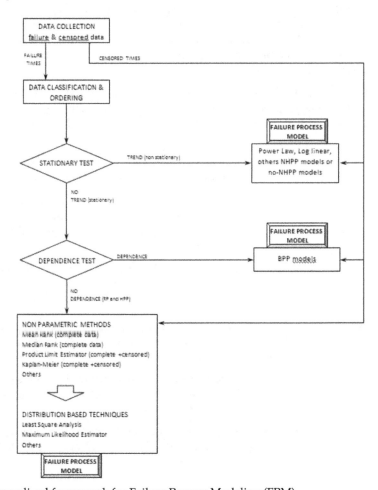

Fig. 1. Generalized framework for Failure Process Modeling (FPM)

In reference to Figure 1, let $X1,j < X2,j < ... < Xi,j < ... < Xn,j$ be the ordered set of failure or inter-failure times of item j; censored times (denoted $Xij+$) are temporarily removed from the data set. The trend test applied to ordered failure times (graphical trend test, Mann test, etc.) determines if the process is stationary or not.

If the process presents a trend, the Xi,j are not identically distributed and a non-stationary model must be fitted. The NHPP model is the most used form due to its simplicity and according to significant experimental evidence available (Coetzee, 1997). At this time the censored data must be reconsidered in the model. Their impact is discussed by Jiang et al. (2005).

If the failure process is trend free, the next step is to identify if inter-failure times are independent. There are a lot of tests for independence, but this check is usually skipped by practitioners, as stated by Ascher and Feingold, because of a lack of understanding of the relevance of this type of test. An effective way of testing the dependence is the serial correlation analysis discussed by Cox and Lewis (1966). Dependence between data involves a Branching Poisson Processes (BPP), which is also analyzed by Cox and Lewis (1966). Censored data also play an important role in the BPP model and must be considered during final modelling.

In real applications the failure process is frquently stationary and the failure data are independent: then a renewal process is involved. In spite of this, the proposed framework pays attention to the evaluation of reliability functions, in particular in presence of censored data.

More precisely, non parametric methods and distribution based techniques are suggested to find the reliability functions such as survival functions, hazard functions, etc. considering censored data.

The Product Limit Estimator method and Kaplan-Meier method for the first category and Least Square Analysis and Maximum Likelihood Estimator technique for the second category are robust and consistent approaches.

Regattieri et al. (2010), Manzini et al. (2009) and Ebeling (2005) discuss in details each method referred in the presented framework.

3. Applications

The proposed framework has been applied in several case studies. In this chapter, two applications are presented, in order to discuss methods, advantages and problems.

The first application deals with an important international manufacturer of oleo-dynamic valves. Using a reliability data set collected during the life of the manufacturing system, the effect of considering or not the censored data is discussed.

The second application involves the application of the complete framework in a light commercial vehicle manufacturing system. In particular, the estimation of the failure time distribution is discussed.

3.1 Application 1: The significant effect of censored data

During the production, the Company collects times to failure using the CMMS system. In particular, the performance of component $r.090.1768$ is analysed; this is a very important electric motor: it is responsible of the movement of the transfer system of the valves assembly line. Figure 2 shows a sketch of the component.

The failure process can be considered as a Renewal Process; stationary test and dependence test are omitted for the sake of simplicity (for details see application 2).

Using non parametric methods, in particular the Kaplan Meyer method (Manzini et al. 2009, Ebeling, 2005) it is possible to evaluate the empirical form of reliability function (called $R(t_i)$).

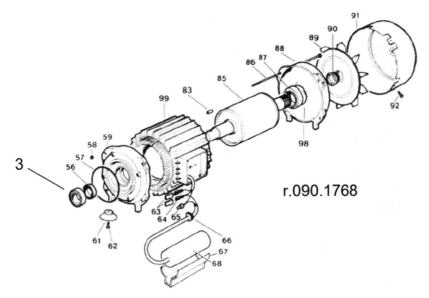

Fig. 2. Component r.090.1768

Table 1 presents all the available data, in terms of times to failures (t_i) and censored times (t_i+)

time to failure (h)	time to failure (h)	time to failure (h)
1124	667	2128
2785	700+	2500+
1642	2756	3467
800+	2489	2687
1974	1500+	1000+
2461	1945	1745
1300+	1478	1000+
2894	1500+	1348
3097	1246	2497
2674	2056	2500+

Table 1. r.090.1768 data set

Assuming ti as the ranked failure times and ni to be the number of components at risk, prior to the ith failure, the estimated reliability is calculated by:

$$\hat{R}(t_i) = \left(1 - \frac{1}{n_i}\right)^{\delta_i} \hat{R}(t_{i-1})$$

(1)

where

$\delta_i = (1,0)$　　　　　(if failure occurs at time ti , if censoring occurs at time ti);

$$\hat{R}(0) = 1$$

The results of reliability analysis are summarized in Table 2 and Figure 3.

i	T_i		n_i	$(1-1/n_i)$	δ_i	R(ti)	
	0						1,000
1	667		30	0.967	1	R(667) = 0.967 R(0) =	0.967
2	700	+	29	0.966	0		
3	800	+	28	0.964	0		
4	1000	+	27	0.963	0		
5	1000	+	26	0.962	0		
6	1124		25	0.960	1	R(1124) = 0.960 R(667) =	0.928
7	1246		24	0.958	1	R(1246) = 0.958 R(1124) =	0.889
8	1300	+	23	0.957	0		
9	1348		22	0.955	1		0.849
10	1478		21	0.952	1		0.808
11	1500	+	20	0.950	0		
12	1500	+	19	0.947	0		
13	1642		18	0.944	1		0.764
14	1745		17	0.941	1		0.719
15	1945		16	0.938	1		0.674
16	1974		15	0.933	1		0.629
17	2056		14	0.929	1		0.584
18	2128		13	0.923	1		0.539
19	2461		12	0.917	1		0.494
20	2489		11	0.909	1		0.449
21	2497		10	0.900	1		0.404
22	2500	+	9	0.889	0		
23	2500	+	8	0.875	0		
24	2674		7	0.857	1		0.346
25	2687		6	0.833	1		0.289
26	2756		5	0.800	1		0.231
27	2785		4	0.750	1		0.173
28	2894		3	0.667	1		0.115
29	3097		2	0.500	1		0.058
30	3467		1	0.000	1		0.000

Table 2. Reliability evaluation using the Kaplan-Meier method (component r.090.1768)

Experimental evidences show that Companies often neglect censored data considering only the times to failure (i.e. the so called *complete* data set).

The use of the complete data set when several components are still working (i.e. there are censored data) introduces significant errors. Considering the component r.090.1768, Figure 4 shows a comparison between the Reliability functions obtained by Kaplan Meyer method applied to the set with censored data, and Improved Direct Method (Manzini et al., 2009) applied only to the failure times (complete data set).

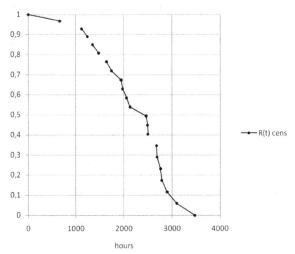

Fig. 3. Reliability Plot using Kaplan Meyer method (component r.090.1768)

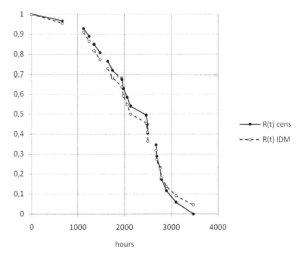

Fig. 4. Comparison of complete (using IDM) and censored data (using Kaplan Meyer)

The error is generally an under estimation of reliability. It depends on the percentage of units not considered and on the censoring times.

Anyway, if censored times are not considered, a significant error is introduced.

3.2 Application 2: A complete failure process modeling

In this application the FPM process has been applied to carry out the reliability analysis of several components of a light commercial vehicles manufacturer production system. The plant is composed by a lot of subsystems; for each of them a set of critical components is considered. Each component has a preferable failure mode (wear, mechanical crash, thermal crash, etc.); from now on, the generic expression *failure* is used considering the particular and prominent failure mode for each component.

Table 3 shows the subset of critical components, called S1, analyzed in the Chapter.

No.	Component	Code	Subsystem
1	tow equipment	KKL5699	welding robot wr1
2	brake shoes	BF4-45	welding robot wr1
3	primary hinge	IHF7598	input gate
4	penstock	KK1243	body feeder
5	skillet ball bearing	A9097	skillet

Table 3. Critical components and subsystems

Among them, the main welding robot named wr1 is very critical, and in particular its components named KKL5699, which are the main actuators, are considered to be mainly responsible for the poor reliability performance of the entire manufacturing system. For this reason, the Chapter presents the application of the proposed framework to the component KKL5699. Finally, the conclusions take into account all the critical components shown in Table 1.

3.2.1 Analysis of KKL5699 component

The tow system is composed by 9 identical repairable components KKL5699, working in 9 different positions, named with letters A,B,..., L, under the same operating conditions. For this reason, they are pooled in a single enhanced data set. The working time is 24 hours/day, 222 days/year.

The CMMS has collected failure data from initial installation ($T_0 = 0$). Table 4 reports the interfailure time X_{ij} (failure i of item j) and the cumulative failure times F_{ij} as shown in Figure 5.

The data are collected during 5 years of operating time, but FPM must be an iterative procedure applied at different instants of system service. The growth of the data set allows a more robust investigation of the failure process. In particular, the paper involves the results of analysis developed at the end of different time intervals $[T_0,t]$: 1,440, 3,696, 4,824, 6,720, 8,448, 11,472, 13,440, 15,560, 18,816 and 23,688 hours. For the generic time t, the analysis uses all the failure times collected, but also the existing censored times according to the components in service. For each instant of analysis, 9 suspended times are collected due to the working times from the last repair action of components and the time analysis. Table 5 reports the data set of failure times, the censored times and the relative working position available at the instant of analysis 3,956 hours.

Fig. 5. Inter-failure time X_{ij} and cumulative failure times F_{ij}

Obviously, when the instant of analysis increases, the number of failure times increases too. Whereas the censored times are constantly 9, then the *Censoring Rate - CR*, given by (1), decreases:

$$CR = \frac{Nc(t)}{Ntot(t)}$$

(2)

where $Nc(t)$ is the number of censored times available at time analysis t, and $Ntot(t)$ is the number of times (failure and censored) available at time analysis t.

The different Censoring Rates involved in the analysis are presented in Table 6.

i	X_{iA} (h)	F_{iA} (h)	X_{iB} (h)	F_{iB} (h)	X_{iC} (h)	F_{iC} (h)	X_{iD} (h)	F_{iD} (h)	X_{iE} (h)	F_{iE} (h)
1	829	829	1.658	1.658	673	673	2.134	2.134	2.724	2.724
2	1.132	1.961	4.536	6.194	983	1.656	1.864	3.998	1.672	4.396
3	2.354	4.315	2.353	8.547	1.567	3.223	3.276	7.274	1.945	6.341
4	1.856	6.171	899	9.446	2.349	5.572	1.745	9.019	2.218	8.559
5	1.957	8.128	1.654	11.100	3.314	8.886	1.794	10.813	1.634	10.193
6	3.420	11.548	1.673	12.773	1.786	10.672	2.398	13.211	2.835	13.028
7	7.264	18.812	2.364	15.137	1.745	12.417	1.128	14.339	1.674	14.702
8	2.573	21.385	1.879	17.016	2.234	14.651	1.523	15.862	1.893	16.595
9	1.111	22.496	2.657	19.673	987	15.638	1.759	17.621	2.673	19.268
10					1.756	17.394	3.298	20.919	1.278	20.546
11					2.567	19.961	1.245	22.164		
12					2.163	22.124				

i	X_{iF} (h)	F_{iF} (h)	X_{iG} (h)	F_{iG} (h)	X_{iH} (h)	F_{iH} (h)	X_{iL} (h)	F_{iL} (h)
1	2.143	2.143	1.980	1.980	3.278	3.278	2.173	2.173
2	1.783	3.926	2.637	4.617	2.167	5.445	2.184	4.357
3	2.784	6.710	1.857	6.474	2.783	8.228	1.476	5.833
4	2.647	9.357	1.643	8.117	1.924	10.152	1.284	7.117
5	1.673	11.030	2.195	10.312	2.735	12.887	749	7.866
6	985	12.015	785	11.097	2.196	15.083	4.583	12.449
7	1.643	13.658	1.903	13.000	2.457	17.540	2.748	15.197
8	1.849	15.507	2.749	15.749	1.932	19.472	2.476	17.673
9	2.649	18.156	2.374	18.123	1.384	20.856	2.945	20.618
10	3.916	22.072	2.734	20.857	1.836	22.692	1.932	22.550
11	673	22.745	1.925	22.782				

Table 4. Inter-failure and failure time data set (KKL5699)

According to the proposed framework, the first test deals with the stationary condition. Figure 6, referring to all the pooled data set, presents the *cumulative failures vs time plot* graphs. No trend can be appreciated in the failure data. The Mann test, counting the number of reverse arrangements, confirms this belief, both for each component and for the pooled data set. The Laplace trend test leads to the same conclusions, in particular its test statistic is $u_L=0.55$ according to a p-value p=0.580. Comparing u_L^2 with the χ^2 distribution with 1 degree of freedom, there is no evidence of a trend with a significance level of 5% (Ansell and Philips, 1994).

interfailure times X_{ij} (h)	position	censored times X_{ij}^{+} (h)	position
829	A	1.735	A
1.132	A	2.038	B
1.658	B	473	C
673	C	1.562	D
983	C	972	E
1.567	C	1.553	F
2.134	D	1.716	G
2.724	E	418	H
2.143	F	1.523	L
1.980	G		
3.278	H		
2.173	L		

Table 5. Data set at 3.956 working hours (inter-failure times and censored times) component KKL5699

Instant of analysis (hs)	1440	2410	3696	4824	6720	8448	11472	13440	16560	18816	23688
N. of interfailure times X_{ij}	2	9	12	18	26	32	46	55	66	75	93
N. of censored times X_{ij}^{+}	9	9	9	9	9	9	9	9	9	9	9
Censoring rate - CR	0,818	0,500	0,429	0,333	0,257	0,220	0,164	0,141	0,120	0,107	0,088

Table 6. Censoring rates (KKL5699)

A first assessment shows that the component failure process is stationary. The next step deals with the renewal process hypothesis evaluation.

The *serial correlation analysis* is used to reveal the independence of the analyzed data set. According to the *autocorrelation plot*, in Figure 7, we cannot reject the null hypothesis of no autocorrelation for any length of lag (5% is the significance level adopted).

The Durbin-Watson statistic confirms this belief, then the component failure process for component KKL5699 can be considered to be a renewal process (RP).

Considering the RP assumption, *non parametric* methods or *distribution based techniques* are available to define the failure process model.

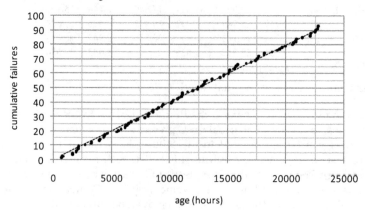

Fig. 6. Cumulative failures vs time plot (pooled data set KKL5699)

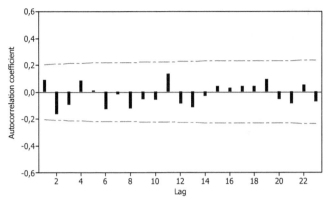

Fig. 7. *Correlogram* for pooled *Xij* with 5% significance level (KKL 5699)

It is important to take into consideration the role of censored data: as demonstrated in the previous application 1, their use enhances the data set and then increases the confidence of the model.

According to the censored data consideration, data are previously analyzed by the *Product Limit Estimator* method (Manzini et al. 2009, Ebeling, 2005), and then the best reliability distributions (i.e. survival function, hazard rate, etc.) are fitted by the *Least square analysis* method. This approach is applied for each time interval of the system service time (i.e. instant of analysis).

From now on, we will only consider the pooled data set; for this reason, the time notation X_{ij} collapses into X_z. The value of survival function - $R(X_z)$, after a working time equal to X_z using the Product Limit Estimator method, is given by:

$$R(X_z) = \left(\frac{n+1-z}{n+2-z}\right)^{\delta_z} R(X_{z-1})$$

(3)

where

$\delta_z = (1;0)$ (if failure occurs at time X_z ; if censoring occurs at time X_z);
$R(0) = 1$;
n number of failure and censored events available

All the service times are investigated. Figure 8 shows the survival empirical curves for component KKL5699, obtained at different instants of analysis, sometimes very spaced out one from each other (e.g. several months apart).

The reliability evaluation after 1,440 hours (roughly 3 months of service) appears very approximate, while after 23,688 hours (more than 4 years of service) it is very confident. Considering the data set, the survival function evaluations change in a significant way (up to 25%) along the life of component; the data collection and the maintenance of data collected is thus a very important issue.

An alternative approach is to identify a proper statistical distribution for the principal reliability functions, such as the survival function, $R(t)$, the failure cumulative probability function, $F(t)$, and the hazard function, $h(t)$, to evaluate its parameter(s), and perform a goodness-of fit test.

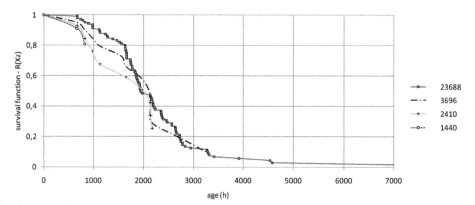

Fig. 8. Survival function at different service times (1,440, 2,410, 3696, 23,688 hours) for component KKL5699

In general, this approach is very interesting because in recent years a significant number of techniques, based on the knowledge of the reliability distributions, have been developed, and then they can provide an important contribution to the performance improvement of both industrial and non-industrial complex systems. These techniques are often referred to as *RAMS (Reliability, Availability, Maintainability and Safety)* analysis.

For example, scientific literature includes a huge number of interesting methods, some being more complex than others, linked to preventive maintenance models; these models support the determination of the best intervention interval, or the optimization of the procedures, that determine spare parts consumption or the best management of their operating costs.

The *2-parameters Weibull* distribution is one of the most commonly used distributions in reliability engineering because of the many failure processes it attains for various values of parameters. It can therefore model a great variety of data and life characteristics. The distribution parameters can be estimated using several methods: the Least Square method (LS), the Maximum Likelihood Estimator (MLE) and others (Ebeling, 2005). In the KKL5699 component case, the Least Square method is preferred for its simplicity and robustness.

Table 7 summarizes the results in terms of Weibull parameters and index of fit, according to different instants of analysis.

Instant of analysis (hs)	1.440	2.410	3.696	4.824	6.720	8.448	11.472	13.440	16.560	18.816	23.688
shape factor β	1,790	2,133	2,489	2,730	2,814	2,749	2,849	2,888	2,989	2,914	2,900
scale factor η	3.160,420	2.087,720	2.268,320	2.251,380	2.378,990	2.393,510	2.311,830	2.335,510	2.335,140	2.400,010	2.427,910
index of fit	0,9451	0,9642	0,9857	0,9871	0,9822	0,985	0,9769	0,9793	0,9788	0,9647	0,9723
Censoring rate - CR	81,8	50,0	42,9	33,3	25,7	22,0	16,4	14,1	12,0	10,7	8,8

Table 7. Parameters of Weibull distribution at different instants of analysis (KKL5699)

The parameters of Weibull distribution move toward steady values of about 3.0 for β and about 2.400 for η. Figure 9 shows graphically the trend of parameters according to different instants of analysis.

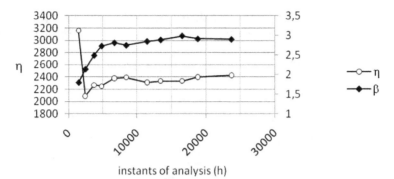

Fig. 9. Trend of Weibull parameters according to the different service times (KKL5699)

Another interesting analysis deals with the link between the estimate of Weibull parameters and the censoring rate. Figures 10 and 11 show the significant role played by CR on β and η paths. The different censoring rates are due to a fixed number of censored data (i.e. 9) and an increasing number of failure data.

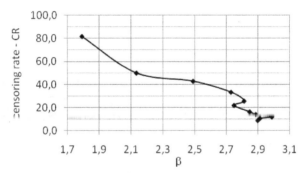

Fig. 10. Shape factor (β) according to the censoring rate (CR) (KKL5699)

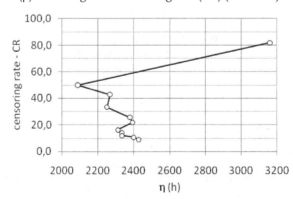

Fig. 11. Scale factor (η) according to the censoring rate (CR) (KKL5699)

Reliability functions, such as the survival function, the cumulative probability of failure and the hazard rate, can be directly derived by 2-parameters Weibull distribution using the estimated parameters of Table 7. Figure 12 shows the survival function R(t) of component KKL5699 for a generic *time interval t* estimated after 2,410, 4,824 and 23,688 hours of service.

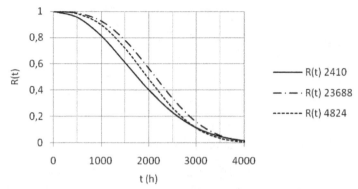

Fig. 12. Survival function (Weibull distribution) according to different instants of analysis (KKL5699)

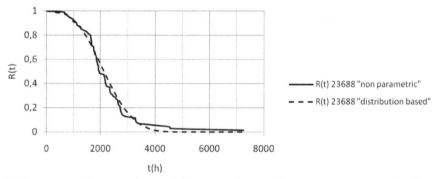

Fig. 13. Comparison between Survival functions obtained by non-parametric method and by 2-parameters Weibull distribution for component KKL5699 after 23,688 hours.

The proposed framework (Fig.1) provides both the *non parametric* and the *distributions based* approaches. Their use usually results in similar outcomes but, as stated before, the second one incorporates more information and is preferable when possible. Figure 13 shows the comparison between the estimates of the survival functions of component KKL5699 obtained after 23,688 hours of service adopting the two different approaches.

3.2.2 Analysis of the subset S1

The stationary and dependence tests of the times to failure highlight how all the components of subset S1 can be described as having a Renewal Process behavior. Failure and censored data are used to perform the Product Limit Estimator method to evaluate empirical values of the cumulative failure distributions. The *2-parameters Weibull* distribution is used to evaluate the reliability functions in an analytical manner; in particular the distribution parameters are estimated using the Least Square method for each instant of analysis. Table 8 summarizes the results in terms of Weibull parameters and in terms of index of fit after 23,688 operating hours.

No.	Component	Code	shape factor β	scale factor η	index of fit
1	tow equipment	KKL5699	2,900	2.427,910	0,9723
2	brake shoes	BF4-45	2,071	3.157,351	0,9885
3	primary hinge	IHF7598	3,541	1.879,467	0,9478
4	penstock	KK1243	2,761	4.578,568	0,9627
5	skillet ball bearing	A9097	2,684	3.125,457	0,9798

Table 8. Parameters of Weibull distribution of S1 components after 23.688 operating hours

Figure 14 shows the survival function of the critical components as calculated on the basis of the parameters reported in Table 8.

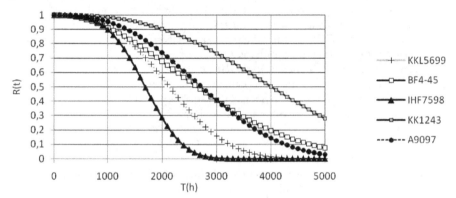

Fig. 14. Survival function of S1 components (2-parameters Weibull)

4. Conclusion

The Failure Process Modeling plays a fundamental role in reliability analysis of manufacturing systems. Complex methodologies are often applied using false assumptions such as constant failure rates, statistical independence between components, renewal processes and others. These misconceptions result in poor evaluation of the real reliability performance of components and systems. All complicated subsequent analysis may be compromised by an incorrect initial assessment relating to the failure mode process.

The experimental evidences show that a correct definition of the model describing the failure mode is a very critical issue and requires efforts often not sufficiently focused on by engineers.

The information collection of both failure data and censored data is a fundamental step. The CMMS method and a system automatically managing the alarms coming from the different sensors installed, can represent valid tools to improve this phase.

As demonstrated in the presented applications, the neglecting of the censored information results in significant errors in the evaluation of the reliability performance of the components.

The knowledge of a fitted analytical distribution is very interesting, because it allows several developments: for example, the determination of the best intervention frequency, or the optimization of the procedures that determine spare parts consumption or the best management of their operating costs.

The FPM procedure must also be maintained: during the service of systems, the reliability data set grows and a more robust estimation of reliability functions is allowed. FPM process, performed using the proposed framework (Fig. 1), must be an iterative procedure renewed during the life of systems.

5. References

Ansell J.I., Phillips M.J., *Practical methods for reliability data analysis*, Oxford University Press, New York, 1994.

Ascher H.E. Feingold H., *Repairable systems reliability. Modeling, inference, misconceptions and their causes*, Marcel Dekker, New York, 1984.

Battini D., Persona A., Regattieri A., Buffer size design linked to reliability performance: a simulative study, *Computers and Industrial Engineering* 56(4), pp. 1633-1641, 2009.

Coetzee J., The role of NHPP models in practical analysis of maintenance failure data, *Reliability Engineering & System Safety*, 56 (2), pp. 161-168, 1997.

Cox D.R., Lewis P.A., *The statistical analysis of series of events*, Methuen, London, 1966.

Ebeling C.E., *Reliability and maintainability engineering*, McGraw-Hill, New York, 2005.

Ferrari E., Gamberi M., Pareschi A., Persona A., Regattieri A., On importance of reliability data collection for failure rate modelling: Reliability Active Data Base – RADB. *9th International Conference on reliability and quality in design*. pp. 132-137. Honolulu – USA, 2003.

Jiang S.T., Landers T.L., Rhoads T.R., Semi-parametric proportional intensity models robustness for right-censored recurrent failure data, *Reliability Engineering & System Safety*, 90 (1), pp. 91-98, 2005.

Louit D.M., Pascual R., Jardine A.K.S., A practical procedure for the selection of time-to-failure models based on the assessment of trends in maintenance data, *Reliability Engineering & System Safety*, Elsevier, London, online.

Manzini R., Regattieri A., Pham H., Ferrari E., *Maintenance for industrial systems*, ISBN: 978-1-84882-574-1, Springer, London, 2009.

Manzini R., Gamberi M., Regattieri A., Persona A., Framework for designing a flexible cellular assembly system, *International journal of production research* 42 (17), pp. 3505-3528, 2004.

Manzini R., Gamberi M., Regattieri A., ,Applying mixed integer programming to the design of a distribution logistic network, *International journal of industrial engineering-theory applications and practice*, 13(2), pp. 207-218. 2006.

Manzini R., Gamberi M., Gebennini E., Regattieri A., An integrated approach to the design and management of a supply chain system, *International journal of advanced manufacturing technology*, 37(5-6), pp.625-640, 2008.

O'Connor P., *Practical reliability engineering*, ISBN:978-0-47084-463-9, Wiley, New York, 2002.

Persona A., Regattieri A., Pham H., Battini D., Remote control and maintenance outsourcing networks and its applications in supply chain management, *Journal of operations management*, 25(6), pp.1275-1291, 2007.

Regattieri A., Gamberi M., Gamberini R., Manzini R., Managing lumpy demand for aircraft spare parts, *Journal of air transport management*, 11(6), pp. 426-431, 2005.

Regattieri A., Manzini R., Battini D., Estimating reliability characteristics in presence of censored data: a case study in a light commercial vehicle manufacturing system, *Reliability Engineering & System Safety*, 95 (10), pp. 1093-1102, 2010.

5

Integrated Process Planning and Scheduling and Multimode Resource Constrained Project Scheduling: Ship Block Assembly Application

Raad Yahya Qassim
Department of Ocean and Naval Engineering, COPPE,
Federal University of Rio de Janeiro
Brazil

1. Introduction

Planning and scheduling are two major tasks in manufacturing system management, which have a direct bearing on the competitive position of enterprises inserted in diverse manufacturing fields, such as the chemical, aerospace, semiconductor, and shipbuilding industries. Inadequate planning and scheduling are considered to be a major cause of the gap between desired and actual enterprise manufacturing performance in terms of inventory level, throughput, manufacturing cost, and facility location. In order to bridge this gap, not only a feasible but also an optimal production network and supply chain schedule is required, which satisfies all types of constraints within the manufacturing – marketing environment, with a view to achieving what has become known as enterprise-wide optimisation (EWO). Until the start of the last decade, the focus of most multiple facility production firms operating at multiple sites has been on operations optimisation at the single facility level. Solutions obtained at this level are suboptimal at the multiple facility level. In order to obtain an optimal solution at the latter level, the scope of optimisation must be enlarged to model the multiple facility supply chain of the enterprise as a whole, including the interaction with suppliers and customers; see for example (Laínez et al., 2010), Munõz et al. , (2011), Stray et al., (2006), and Monostori et al., (2010).

Within EWO, there exists a wide spectrum of optimisation problems whose nature depends on the type of manufacturing environment that is under consideration. The concern in this paper is focused on two such problems:

1. integrated process planning and scheduling problem (IPPSP);
2. multimode resource-constrained project scheduling problem (MRCPSP).

In view of their important role in manufacturing management practice, this pair of problems has attracted significant interest in the academic literature. In the IPPSP, the two functions of process planning and production scheduling, which have usually been solved sequentially, are integrated and solved simultaneously, with a view to providing greater flexibility at the factory floor level. In the MRCPSP, the schedule of a project is obtained allowing each activity one or more modes of execution, whereby with each mode a time duration, execution cost, and resource consumption level are associated. For recent surveys of the IPPSP and MRCPSP, the reader may refer to (Shen *et al.*, 2006) and (Hartmann & Briskorn, 2010), respectively. The IPPSP arises normally at the shop floor level in batch process and discrete part manufacturing environments. The MRCPSP occurs in project – oriented production environments, such as those in shipbuilding, aerospace, and highway construction; see (Martínez *et al.*, 1997 and Martínez & Pérez, 1998). The general objective of this paper is to relate these two hitherto separately considered problems in manufacturing system management, in order to open vistas for the formulation and solution of new problem variants, which may benefit from the employment of methods and techniques that have been already developed separately for the IPPSP and MRCPSP. A specific application is provided by the development of a mathematical programming model for ship block assembly, which is one of the major final stages in shipbuilding; see for example (Yu-guang *et al.*, 2011).

1.1 Integrated process planning and production scheduling problem

In discrete manufacturing system management, two major tasks are process planning and production scheduling. In process planning, the question of how an item will be manufactured is answered. This is done by the determination of the sequence of operations that are necessary to produce the item under consideration. An item may be a part or an assembly of several parts. In general, an item may be manufactured in number of ways. This leads to the existence of a multiplicity of process plans. On the other hand, the question that has to be answered in production scheduling is the following: when is a job be dispatched to the shop floor, and what amount of each resource is allocated for its manufacture under prevailing conditions at a specific instant of time?

In general, process planning consists of the determination of operations and parameters that are required to convert raw materials or intermediate items into a finished item, such as a part or an assembly. The process planning task comprises the interpretation of design data, selection and sequencing of operations to manufacture the item, selection of machines, tools, and other resources, along with the corresponding quantitative data such as machine speeds and resource amounts. It is common practice that process planning is carried out in two stages: preliminary and detailed, whereby feasible process plans and optimal process plans are generated, respectively. Specific aspects of process planning vary with the application environment, such as machining, welding, and assembly; see, for example (Kong *et al.*, 2011).

Production scheduling is normally carried out for several time horizon spans and corresponding levels of detail. This approach is known as hierarchical production planning (HPP); see, for example, (Bang & Kim 2010). At the shortest time horizon span with corresponding most extensive level of detail, stands the shop floor scheduling problem, which is the focus of this paper. Here, one has a set of jobs, each of which possesses its own set of process plans, along with available resource amounts at the shop floor level. One requires the sequencing of jobs in time and the corresponding allocation of resources to each job; see, for example, (Li, *et al.*, 2010).

In practice, the tasks of process planning and production scheduling are carried out sequentially, whereby for each item a set of alternative process plans are first determined. The generation of each process plan of an item is based on the assumption of the unlimited availability of the set of resources that are necessary for the execution of the plan on the shop floor. For each operation within a process plan, an execution time and an amount of each resource that is required for its execution, are associated. In the sequential approach, one member of the set of process plans so generated is selected according to one or more criteria, such as execution cost and execution time. Once the process plan selection problem is solved, and the process plan is fixed for each item, the production scheduling task is carried out for a set of jobs, each of which corresponds to an item, employing available resource amounts on the shop floor. This decomposition of the overall problem into a sequence of two subproblems for the generation of process plans and production schedules defines the sequential approach.

The sequential approach possesses the undeniable advantage of the simplification by decomposition of a large problem into a pair of smaller subproblems; however, the decomposition constitutes in essence a heuristic, in that the solution that it provides is an approximation to the solution of the original process planning / production scheduling problem. As a heuristic, it may provide adequate solutions albeit not optimal. In fact, an optimal solution can only be envisaged if a simultaneous approach is adopted, whereby the original problem is attacked without its decomposition into a pair of subproblems. This has become to be known as the integrated process planning and scheduling (IPPSP). Clearly, the simultaneous approach, whilst providing an optimal solution, involves a larger and therefore more difficult problem than that involved in the sequential approach. An additional and important practical advantage of the simultaneous approach is that under certain conditions in the shop floor, it may provide feasible solutions when this is not possible with the sequential approach. An example of this arises when job due dates are excessively tight when process plans are fixed prior to job arrival on the shop floor, as is the case in the sequential approach. In contrast, in the simultaneous approach, the degree of freedom provided by selecting process plans along with job production scheduling decisions may allow for meeting job due dates.

Having provided a motivation and a justification for research efforts on the IPPSP, the next question that arises pertains to its formulation and solution. A start is made by providing a typical problem statement of the IPPSP as follows. Given a set of independent jobs, with each of which a delivery due date is associated, that is to be manufactured employing one of a set of process plans, each of which consisting of a sequence of operations, with each of which a processing time and a set of resource amounts is associated, it is required to determine a minimum makespan production schedule. Clearly, in the IPPSP what is being sought is an optimal schedule. Consequently, mathematical programming (MP) provides a natural framework for the formulation and solution of the IPPS problem. Extensive work has been reported on variants of the IPPSP employing MP models; see (Li, et al., 2010) for further details.

1.2 Multimode resource constrained project scheduling

A major task in project management is project scheduling whose objective is the sequencing of project activities, which may be executed in one or more modes, subject to logical precedence between activities and limited resources. This has given rise to the resource – constrained project scheduling problem (RCPSP). Extensive work has been carried out on the RCPSP; see [5] for a recent literature review. An important extension of the RCPSP is the multimode variant (MRCPSP), whereby an activity may be executed in more than one mode, whereby mode is a proxy for intensity, in that the activity work content may be executed at various speeds and corresponding resource consumption rates.

A typical problem statement of the MRCPSP may be provided as follows. Given a project that consists of a set of activities with a corresponding set of logical precedence relations, a set pair of renewable and nonrenewable resources, it is required to determine a minimum makespan schedule. Obviously, the goal is the generation of an optimal schedule, and for this mathematical programming (MP) constitutes a natural framework for the formulation and solution of the MRCPSP. An extensive literature exists on MP modeling of the MRCPSP; see (Hartmann & Briskorn, 2010), for a recent review.

2. Process plan – Execution mode relation

The similarity between two a priori different problems that arise in manufacturing systems management, IPPSP and MRCPSP, stems from the analogous roles of process plan and activity mode in the IPPSP and MRCPSP tasks, respectively. A job that is to be produced may be viewed as an activity to be completed, whilst a process plan corresponds to an execution mode. Extending the analogy, multiple jobs correspond to multiple projects in the corresponding environment.

If on the one hand, a process plan is similar to an execution mode, there exists an important conceptual difference. The former consists of a set of operations which possesses a corresponding set of precedence relations, whilst the latter is indivisible. Nevertheless, this conceptual difference results in practical implications when MP models are formulated and solved for the IPPSP and MRCPSP. In particular, the number of constraints in the former is higher in the IPPSP than in the MRCPSP of equal size, due to the existence of precedence relations in a process plan as opposed to the indivisibility of an execution mode. On a closer look, this difference may be conveniently removed if one considers the division of activities into subactivities, as suggested in (Nicoletti & Nicoló, 1998). The introduction of the concept of subactivity may not be merely semantic, as it may possess a practical aspect, such as activity preemption. The interruption of activities has been shown to provide a manner of fast tracking. In practice, an activity may only be interrupted at a finite number of points during its execution. These points may serve as a useful basis for the definition of subactivities.

3. Ship block assembly

Normally, the assembly of a ship block is part of a shipbuilding project, which comprises several blocks. As a result, the assembly of each block should be appropriately scheduled, so as to be compatible with the schedule of the shipbuilding project as a whole. In general, it is desirable to determine the earliest delivery date of each block. The problem of interest then is to determine the minimum makespan assembly schedule for each block. This problem may be viewed as a project scheduling problem with resource and material supply constraints; see, for example, Dodin & Elimam, 2001, Alfieri et al., 2010, and Sajadieh, et al., 2009.

The following assumptions are made with a view to facilitating modelling of the problem.

1.   With each activity, a duration time and a set of direct predecessor activities, and a set of direct successor activities are associated.
2.   With each activity, a set of non – renewable resources, these being materials, is associated, this set being necessary to start and finish the activity.
3.   With each member of the material set, a procurement lead time and a spatial area are associated, this set being necessary to start and finish the activity.

4. With each activity, a set of renewable resources, such as spatial area and manpower, is associated.

Spatial area, which is a member of the set of renewable resources, is employed for two purposes: execution of an activity and storage of associated materials from delivery time to the start time of the execution of the activity.

The following notation is employed:

i, j – indices for activities; $i \in I$

k – index for non-spatial area renewable resource; $k \in K$

m – index for non-renewable material source; $m \in M$

n – index for spatial area renewable resource; $n=1$

t – index for time period; $t \in T$

D_i – duration time of activity i

EF_i – earliest finish time of activity i

ES_i – earliest start time of activity i

F_i – set of direct predecessor activities of activity i

G_m – lead time material m

H_{imn} - spatial area renewable resource n required by material m associated with activity i

LF_i – latest finish time of activity i

LS_i – latest start time of activity i

P_{ik} – non-spatial area renewable resource per unit time required by activity i

Q_n – available non - spatial area renewable resource n

R_{mt} – inventory level of material non – renewable resource m in time period t

S_n – available spatial area renewable resource n

X_{it} – binary variable $\in$ { 1 if activity i is finished in time period t, 0 otherwise }

Y_{it} – binary variable $\in$ { 1 if activity I is in execution in time period t, 0 otherwise }

Z_{imt} – binary variable $\in$ { 1 if material m associated with activity I is ordered in time period t, 0 otherwise }

The model may be stated as follows.

$$\text{Minimise } \Sigma_{\tau=EFI,...LFI} \ \tau \ X_{I\tau} \ , \tag{1}$$

subject to

$$\Sigma_{\tau=EFj,...,LFj} \ \tau \ X_{j\tau} + D_i \leq \Sigma_{\tau=EFi,...,LFi} \ \tau \ X_{i\tau} \ , \quad \forall j \in F_i \ ; \ \forall i \in I \ , \tag{2}$$

$$\sum_{\tau=1,...,T} \tau \ Z_{im\tau} \leq \sum_{\tau=EFi,...,LFi} \tau \ X_{i\tau} - D_i - G_m \ , \forall i \in I \ ; \ \forall m \in M \ , \tag{3}$$

$$\Sigma_{i \in I} \ P_{ik} \ Y_{it} \leq Q_k \ , \quad \forall k \in K \ ; \ \forall t \in T \ , \tag{4}$$

$$\Sigma_{m \in M} \quad R_{mt} \leq S_n \ , \quad \forall t \in T \ , \tag{5}$$

$$R_{mt} = R_{m,t-1} + \Sigma_{i \in I} \ H_{imn} \ (\ Z_{im,t-Gm} - X_{it}) \ , \forall m \in M \ ; \ \forall t \in T \ , \tag{6}$$

$$R_{m0} = 0 \ , \forall m \in M \ , \tag{7}$$

$$\Sigma_{\tau=EFi,...,LFi} \ X_{i\tau} = 1 \ , \forall i \in I \ , \tag{8}$$

$$\Sigma_{\tau=EFi,...,LFi} \ Y_{i\tau} = D_i \ , \ \forall i \in I \ , \tag{9}$$

$$\Sigma_{\tau=1,...,T} \ Z_{im\tau} = 1 \ , \ \forall i \in I \ ; \ \forall m \in M \ , \tag{10}$$

$$X_{it} , Y_{it} , Z_{imt} \in \{ \, 0 \, , 1 \, \} \ , \ \forall i \in I \ ; \ \forall m \in M \ ; \ \forall t \in T \ . \tag{11}$$

Expression (1) defines the project makespan objective function that is to be minimised. Constraints (2) ensure precedence between an activity and each of its direct predecessor activities. Constraints (3) ensure that all materials necessary for each activity are available before the start of an activity. Constraints (4) and (5) guarantee that each renewable resource is not exceeded in each time period. Constraints (6) ensure the material balance over the planning horizon. Constraints (7) fix the initial inventory level of each material at the start of the planning horizon. Constraints (8) ensure that an activity is finished once in the planning horizon. Constraints (9) define the time duration of each activity. Constraints (10) guarantee that each material is ordered once for each activity over the planning horizon. Constraints (11) define the domain of the decision variables.

3.1 Three – level assembly problem

Consider for illustrative purposes the three – level block assembly example shown in Fig.1. This problem possesses features which belong to both the IPPSP and MRCPSP. As it arises in shipbuilding practice, it is naturally viewed as a variant of the IPPSP. With a view to highlighting the connection with the MRCPSP, it is modelled in this Subsection as belonging to the MRCPSP. The initial and final activities are denoted by i= 0 and i= 8, respectively. It is assumed that there exists a single non-spatial renewable resource, namely manpower, and it is denoted by k=1; furthermore, it is assumed that the spatial area renewable resource is common to all activities and materials in the assembly workshop and it is denoted by n=1. The model of this example may be stated as follows.

$$\text{Minimise } \Sigma_{\tau=EF8,...,LFT8} \ \tau \, X_{8\tau} \ , \tag{12}$$

subject to

$$\Sigma_{\tau=EF0,...,LF0} \ \tau \, X_{0\tau} + D_1 \le \Sigma_{\tau=EF1,...,LF1} \ \tau \, X_{1\tau} \ , \tag{13}$$

$$\Sigma_{\tau=1,...,T} \ \tau \, Z_{11\tau} \le \tau X_{1\tau} - D_1 - G_1 \ , \tag{14}$$

$$\Sigma_{i=1,...,8} \ P_{i1} \ Y_{1t} \le Q_1 \ , \tag{15}$$

$$\Sigma_{m=1,...,8} \ R_{m1} \le S_1 \ , \tag{16}$$

$$R_{12} = R_{11} + \Sigma_{i=1,...,8} \ H_{i11} \ (\, Z_{i1,1 \, -G1} - X_{11} \,) \ , \tag{17}$$

$$R_{10} = 0 \ , \tag{18}$$

$$\Sigma_{\tau=EF1,...,LF1} \ X_{1\tau} = 1 \ , \tag{19}$$

$$\Sigma_{\tau=EF1,...,LF1} \ Y_{1\tau} = D_1 \ , \tag{20}$$

$$\Sigma_{\tau=EF1,...,LF8} \ Z_{11\tau} \ . \tag{21}$$

Integrated Process Planning and Scheduling and Multimode Resource Constrained Project Scheduling:
Ship Block Assembly Application

83

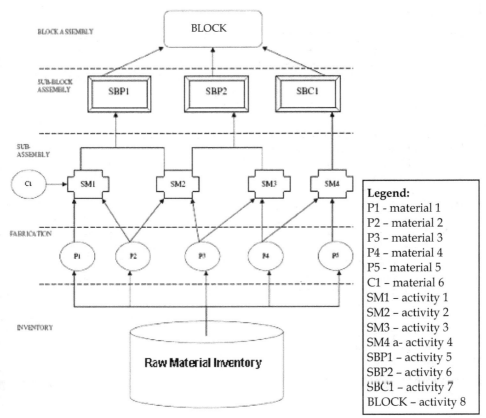

Fig. 1. Three – level assembly example.

4. Conclusions

In this paper, the relation between two problems, IPSS and MRCPSP, arising in manufacturing systems management, which have been formulated and solved by separate mathematical programming modeling approaches, have been shown to possess clear similarity features. This similarity has not been explored, and therefore there exists a clear potential for the interplay of methods and techniques between the two problems. These include novel model formulations and solution strategies for the IPPSP and MRCPSP variants, such as dynamic shop floor scheduling and project rescheduling (Ouelhadj & Petrovic, 2009), (Gerk & Qassim, 2008). This provides a rich field to be explored by future research work, a start having been made in a recent paper (Capek et al., 2011).

5. Acknowledgements

The author wishes to acknowledge financial support, in the form of a research fellowship, of the Brazilian National Research and Development council, CNPq. Recommendations and suggestions by an anonymous referee with a view to improving the paper are gratefully acknowledged.

6. References

Alfieri, A., Tolio, T., Urgo, M. (2010) A Project scheduling approach to production and material requirement planning in Manufacturing-to-Order environments. *Journal of Intelligent Manufacturing*, DOI 10.1007/s10845-010-0396-1.

Bang, J.-Y.; Kim, Y.-D. (2010). Hierarchical production planning for semiconductor water fabrication based on linear programming and discrete – event simulation. *IEEE Transactions on Automation Science and Engineering*, vol.7, no.2, pp,326-336.

Capek, R.; Sucha, P.; Hanzálek, Z. (2011). Production scheduling with alternative process plans. *European Journal of Operational Research*, doi: 10.1016/j.ejor.2011.09.018.

Dodin, B., Elimam, A.A. (2001) Integrated project scheduling and material planning with variable activity duration and rewards. *IIE Transactions*, vol.33, pp.1005-1018.

Gerk, J. E. V.; Qassim, R. Y. (2008) Project acceleration via activity crashing, overlapping, and substitution. *IEEE Transactions on Engineering Management*, vol.55, no.4, pp. 590-601.

Hartmann, S.; Briskorn, D. (2010). A survey of variants and extensions of the resource - constrained project scheduling problem. *European Journal of Operational Research*, vol.207, pp.1-14.

Kong, F.; Zuo, J.; Zha, G.; Zhang, J. (2011). Study and application of open die forging CAPP system based on process knowledge. *Journal of Advanced Manufacturing Systems*, vol.10, no.1, pp.45-52.

Laínez, J.M.; Reklaitis, G.V.; Puigjaner, L. (2010). Linking marketing and supply chain models for improved business strategic decision support. *Computers and Chemical Engineering*, vol.34, pp.2107-2117.

Li, X.; Shao, X.; Zhang, C.; Wang, C. (2010). Mathematical modeling and evolutionary – based approach for integrated process planning and scheduling. *Computers and Operations Research*, vol.37, pp.656-667.

Martínez, E. C., Duje, D., Pérez, G.A. (1997) On performance modeling of Project-oriented production. *Computers and Industrial Engineering*, vol.33, no.3, pp.509-527.

Martínez, E. C., Pérez, G. A. (1998) A project-oriented production model of batch plants. *Computers and Chemical Engineering*, vol.22, no.3, pp.391-414.

Monostori, L., Edos, G., Kádár, B., Kis, T., Kovács, A., Pffeifer, A., Váncza, A. (2010) digital enterprise solution for integrated production planning and control. *Computers in Industry*, vol.61, pp.112-126.

Munõz, E.; Capón-Garcia, E.; Moreno-Benito, M.; Espuña, A.; Puigjaner, L. (2011). Scheduling and control decision-making under an integrated information environment. *Computers and Chemical Engineering*, vol.35, pp.774-786.

Nicoletti, S., Nicoló, F. (1998). A concurrent engineering decision model : Management of the project activities information flow. *International Journal of Production Economics*, vol.54, pp.115-127.

Ouelhadj, D.; Petrovic, S. (2009). A survey of dynamic scheduling in manufacturing systems. *Journal of Scheduling*, vol.12, pp.417-431.

Sajadieh, M.S., Shadrkh, S., Hassanzadeh, F. (2009) Concurrent project scheduling and material planning: A genetic algorithm approach. *Scientia Iranica*, vol. 16, pp.91-99.

Shen, W.; Wang, L.; Hao, Q. (2006). Agent-based distributed manufacturing process planning and scheduling: A state – of – the – art survey. *IEEE Transactions on Systems, Man, and Cybernetics – Partt C: Applications and Reviews*, vol.36, no.4, pp.563-577.

Stray, J.; Fowler, J.W.; Carlyle, W.M.; Rastogi, A.P. (2006). Enterprise-wide semiconductor resource planning. *IEEE Transactions on Semiconductor Manufacturing*, vol.19, no.2, pp.259-268.

Yu-guang, Z., Kai, X., Yong, Z. (2011). Modeling and analysis of panel hull block assembly system through time colored petri net. *Marine Structures*, vol.24, no.4, pp.570-580.

6

Migrating from Manual to Automated Assembly of a Product Family: Procedural Guidelines and a Case Study

Michael A. Saliba* and Anthony Caruana
Department of Industrial and Manufacturing Engineering, University of Malta
Malta

1. Introduction

A challenge that is often faced by product manufacturers is that of migration from a manual to an automated production system. The need to embark on this migration may develop from a number of different scenarios. Typical triggers for the automation of a manual production process include a need to increase competitiveness (by reducing labour costs and increasing labour productivity), a need to meet higher quality demands from the customer, an increased awareness of health and safety issues leading to the need to move human workers away from hazardous tasks, and a need to improve production efficiency parameters (e.g. reduce manufacturing lead time, improve production capacity, improve production flexibility and agility). The general strategic and/or technical aspects pertaining to the implementation of automation have been widely addressed in the literature (e.g. Asfahl, 1992; Chan & Abhary, 1996; Groover, 2001; Säfsten et al., 2007). In this work the focus is on the compilation and application of a number of standard design tools and of production system evaluation tools to facilitate and support the migration to an automated production system, in a scenario that involves a certain degree of product variety. The results are presented in the form of a set of recommended procedural guidelines for the development of a conceptual solution to the migration process, and the implementation of the guidelines in a real industrial case study.

Specifically, this work addresses the situation where a manufacturer needs to investigate a potential manufacturing system migration for the assembly of a part family of products, where no or minimal product design changes are allowed. A list of procedural guidelines is proposed, in order to aid the manufacturer in analyzing the requirements for the transition, and in carrying out a conceptual design for a suitable automated manufacturing system. The guidelines are applied in the context of the industrial case study, where it is required to investigate and develop a migration plan for the assembly of three related product part families.

The case study involves a relatively large manufacturing plant (about 700 employees), that produces electromechanical switch assemblies for the automobile industry. The trigger for

---

* Corresponding Author

the migration process is a significant increase in projected order volumes over a four year period, thus necessitating an increase in production capacity, as well as providing a good opportunity to obtain substantial return on investment following the implementation of advanced manufacturing technologies. The three part families under consideration are referred to as the "single gang" switches consisting of 20 variants, the "old three gang" switches consisting of 11 variants, and the "new three gang" switches consisting of 9 variants. A representative member of each of these families is shown in the illustration in Figure 1. Switch assembly is currently carried out in a mainly manual manner, with the aid of pneumatic presses to provide the required clipping forces. The projected production volumes for the current year (Year 1) and for the subsequent years are summarized in Table 1.

The academic goals of this research are (i) to compile a set of guidelines for migration in a scenario of this type, and to apply the guidelines to this case study; (ii) to perform a study on assembly related similarities of the product families and to take advantage of these similarities in the automation process; (iii) to interpret the analytical results obtained from the feasibility analysis, so as to define the most suitable assembly line; and (iv) to utilize analytical tools in order to define the best possible concept at all stages of the automation. From an industrial perspective, the goals are (i) to perform a cost reduction exercise on the current assembly processes; (ii) to perform a feasibility analysis of the developed concepts with respect to a number of considerations such as the assembly line balancing, cycle time reduction, production capacity and maintenance requirements; and (iii) to plan the integration of ergonomic principles in all workstations and also develop and implement safety guidelines in the process development.

Single Gang   **Old Three Gang**   **New Three Gang**

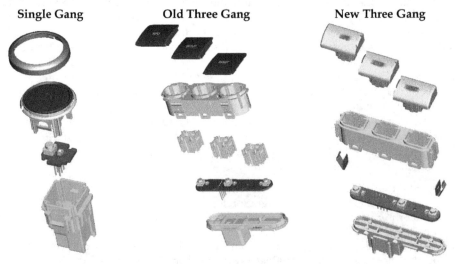

Fig. 1. An exploded view of a representative member of each product family

	Year 1	Year 2	Year 3	Year 4
Single gang switches	136,000	935,000	1,432,600	1,552,600
Old 3 gang switches	1,098,000	816,000	816,000	816,000
New 3 gang switches	0	150,000	865,000	1,495,000

Table 1. Projected annual production volumes

2. Literature review

In product manufacture, assembly needs to be given significant importance, since assembly related operations generally amount to 70% of the total product cost (Boothroyd et al., 2001). Therefore rationalisation of the assembly process is essential in order to optimize, mechanize and automate the activities performed, especially for assembly-intensive products. Where a new product is being developed, an *Integrated Product Development* (IPD) approach can be taken, whereby the marketing, design, and manufacture of the product family can be optimized and developed concurrently (Andreasen & Hein, 2000). Where the product family exists and is already in manual production, the options for design modifications may be minimal, and the development of the automated production system rests on effective categorization of the products through analytical methodologies such as group technology (e.g. Hyer & Wemmerlov, 2002), and the effective exploitation of the identified similarities.

Several approaches are found in the literature to address the migration problem. Asfahl (1992) identified five phases in the implementation of automation to a currently manual process: *planning, development, mock-up and test, installation,* and *production and follow-through.* He further highlighted a number of key activities to be carried out during the planning phase: the isolation of the potential application; the identification of the project objectives; the consideration of the drawbacks; the early planning of safety aspects; the detailed documentation of the current (manual) operation; the selection of fixed versus flexible automation; and the development of a proposed layout for the system. Chan and Abhary (1996) applied an analytic hierarchy process to compare three different potential automation strategies to the existing manual plant in a case study, using a simulation approach and several evaluation criteria. Kapp (1997) introduced the "USA Principle" – Understand the existing process, Simplify the process, and Automate the process, originally intended as a guide for the implementation of enterprise resource planning (ERP), but applicable as a straightforward approach to all automation projects. Groover (2001) suggested a three-phase process whereby the individual processing stations are automated first, followed by the integration of the systems through automated handling between stations. Baines (2004) recommended a nine step approach to the manufacturing technology acquisition process: technology profiling; establishment of technology requirements; identification of a technology solution; formation of an outline business case; selection of a technology source (which may include internal development of the technology); demonstration of the technology; confirmation of the business case; implementation of the technology; and post-investment audit. Säfsten et al. (2007) suggested that manufacturing strategy development could be based on function allocation, employing a system design process that allocates various functions to either humans or machines, and optimizing the level of automation in the plant. Winroth and Säfsten (2008) further suggested an automation strategy whereby the bottom-up activities (stemming from the internal need for improvement) and the top-down activities (stemming from the market requirements) are both taken into consideration in the optimization exercise.

Most of the technical literature on the development of automation systems focuses on the process that needs to be automated and/or on the product that needs to be manufactured, but less so on the production system as a product in its own right. Thus, a number of design methods that have become a mainstay in product design are not normally prescribed in a systematic way to the development of integrated automation systems. Design methods include tools and techniques such as product design specification (PDS), morphological charts, decision matrices, and failure modes and effects analyses (FMEA) (e.g. Dieter & Schmidt, 2009). This is the research gap that has been identified in this work, and the results reported in this chapter attempt to bridge this gap by drawing on these design methodologies, and prescribing them side by side with other conventional developmental steps, in order to optimize the conceptual design process for an automation system, in an environment of variety in the products that need to be manufactured. The contribution of this work is further extended to include a detailed illustration of the step by step application of the procedural guidelines to a complex industrial case study.

3. The procedural guidelines

A systematic approach to the conceptual design of a new, automated manufacturing system when migrating from manual assembly of a product part family, where little or no change in product design is allowed, is presented in Table 2. The proposed list of procedural guidelines and development tools given in the table has been compiled on the general basis of the discussion given in section 2, and the developmental steps are intended to be applied sequentially, for the case of a high production volume environment in the presence of product variety. The guidelines are intended to cover the early technical and feasibility studies. Thus it is pre-assumed that the company has already taken a strategic decision to analyze the selected manual process with a view to implementing automation (if feasible), and the guidelines lead to the end of the conceptual design phase but do not address any part of the embodiment design phase or of the development of test or prototype hardware. In the general literature, it has been estimated that about 75% of the product cost is normally already committed by the end of the conceptual design stage (Ullman, 1997), and in this work the research boundary has been set to address this critical phase of product development. It is emphasized once again that it is the *production equipment* that is being referred to and considered as the "product" in the context of the previous sentence, rather than the objects (products, or product part family) that will be manufactured by the equipment.

In the following section, the use and implementation of the procedural guidelines is illustrated in the context of the industrial case study, for the development of the conceptual design of a new manufacturing system for the assembly of the three families of automotive switches. The results for each step are summarized, presented and discussed.

4. Implementation of the procedural guidelines: A case study

4.1 Analysis of the product family designs

In a manufacturing environment, parts having similar geometric shapes and sizes or similar processing steps may be grouped into part families, in order to facilitate their design and/or their production. This manufacturing philosophy is referred to as *group technology*. Thus, parts within a particular part family will all be uniquely different, however they will have enough similarities to classify them together as one group (e.g. Groover, 2001). In the present case study, the product designs are fixed, and therefore the application of group technology principles is intended to facilitate the production of the parts. Parts classification systems are normally based either on similarities in design attributes, or on similarities in manufacturing attributes, or on similarities in both design and manufacturing attributes; however other types of similarity may also be used. Hyer and Wemmerlov (2002) identify nine criteria that may be used to classify parts, based on similarities in product type, market, customers, degree of customer contact, volume range, order stream, competitive basis, process type, and/or product characteristics.

1.	Analyze the current product family design, with a view to understanding clearly all similarities and variations between the members of the family and between their components.
2.	Analyze the current assembly processes and the existing assembly line(s), with a view to understanding the processes, and identifying drawbacks and opportunities for simplification.
3.	Perform a capacity analysis based on the current set-up, with a view to understanding and defining current capabilities and limitations.
4.	Draw up a product design specification (PDS) chart for the new production system, with a view to defining the requirements and wishes for the new system.
5.	Perform a group technology (GT) analysis, with a view to confirming/revising the parts classification in the context of automated manufacture.
6.	Create precedence diagrams for the process, with a view to understanding the various ways in which assembly operations can be carried out.
7.	Set up a morphological chart for the overall operation, with a view to identifying various alternatives for carrying out the various process steps.
8.	Draw up a number of different layouts at the conceptual level, with a view to identifying different alternatives for the assembly.
9.	Perform a provisional analytical study of each of the concepts, based on various criteria such as achievable cycle times, quality, shop floor area, and flexibility.
10.	Draw up a decision matrix to select the most suitable concept.
11.	Carry out a process failure modes and effects analysis (PFMEA), with a view to identifying and addressing failure mechanisms.
12.	Perform a safety analysis, with a view to identifying and addressing production hazards.
13.	Perform an ergonomic analysis, with a view to optimizing the production system with respect to interactions with human workers.
14.	Carry out a new capacity analysis for the new system, with a view to quantifying the achievable capabilities through automation.
15.	Perform a provisional return on investment analysis, with a view to quantifying provisionally the projected savings and break even times upon implementation of the new system.

Table 2. The procedural guidelines

In this case study, a preliminary analysis based on product design strongly indicated that the parts fell into three natural groupings as shown in Figure 1 above, based on the overall features of their geometries. All of the 20 variants of single gang switches included a socket, a printed circuit board (PCB), and a push button, as shown in Figure 2(a). The PCB included one or more coloured light emitting diodes (LEDs) as required. Some of these variants included one or more of three additional parts: a chrome ring to provide a different aesthetic finish (such as in the switch shown in Figure 2(b)), a light shield for variants that had two different graphics on their front face (to prevent light leakage between graphics), and a jewel

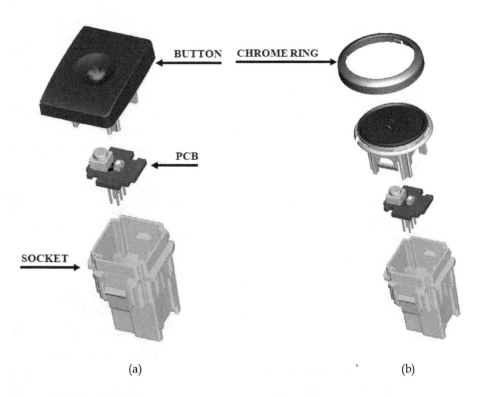

(a)   (b)

Fig. 2. Single gang switches. (a) a basic variant, (b) variant with a chrome ring

(press fit into the button) to transmit light from the LED to the surface of the button. All of the 11 variants of old three gang switches included a socket, a PCB, one or more sliders, a three unit housing, and three buttons as shown in Figure 3(a). The differences between the variants were defined by the types and combinations of buttons (functional, display, or blank). Functional buttons require a slider, in order to actuate a tact switch on the PCB, and also have a graphic display. Display buttons have only a graphic display (illuminated by an LED on the PCB), and blank buttons have no function or display. The nine variants of new three gang switches have a sleeker design, and use metal clips to attach to the dashboard of the vehicle (see Figure 3(b)).

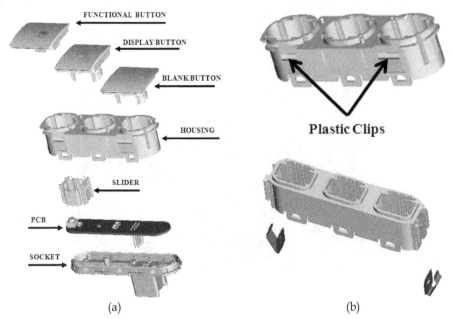

Fig. 3. (a) A variant of old three gang switch, (b) clipping mechanisms (top – old three gang; bottom – new three gang)

4.2 Analysis of the current manufacturing processes

The layouts of the existing production lines are illustrated in Figure 4, with the old three gang switches and some of the single gang switches manufactured in Cell 1, and the new three gang switches and the rest of the single gang switches manufactured in Cell 2. Due to limitations in the end of line testing steps, only one model of switch can be assembled on each cell at any one time, and substantial set-up times are associated with the change over between batches of different models of switch.

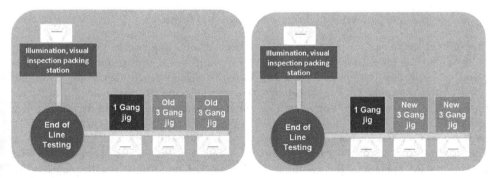

Fig. 4. (a) Layout of Cell 1, (b) layout of Cell 2

The labour intensive nature of the assembly process for the single gang switches is illustrated in Figure 5 and Figure 6. The operator reaches for one socket from the silo, and one PCB from the tray and places the socket over the PCB in cavity (1). If a chrome ring or light shield is required for the switch being assembled, the operator reaches for the chrome

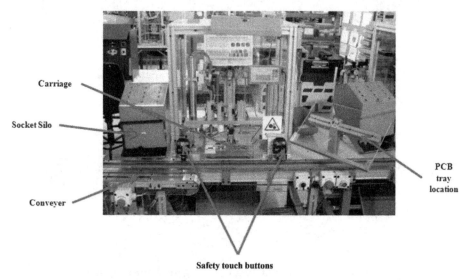

Fig. 5. Assembly jig for single gang switches

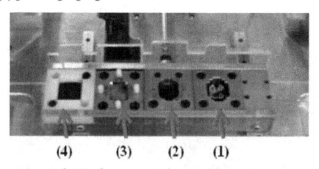

Fig. 6. The four cavities on the single gang switch assembly carriage

ring or light shield and for the button and performs manual alignment. A chrome ring and button sub-assembly is placed in cavity (3), while a light shield and button sub-assembly is placed in cavity (4). The operator then presses the two safety touch buttons placed on the sides of the assembly jig simultaneously. The carriage moves inside the jig and two pneumatic cylinders clip the socket and button sub-assemblies. Subsequently, the carriage moves outside the jig and the operator picks up the socket sub-assembly from cavity (1), rotates it and places it inside cavity (2), whilst placing the button sub-assembly on top of it. The operator also loads the parts for another socket and button sub-assemblies so that during the jig operation, three clipping processes are performed simultaneously. The operator then presses the two safety buttons simultaneously and the carriage moves inside the jig and a pneumatic cylinder clips the two sub-assemblies together. When the carriage moves outside the jig, the operator removes the switch and places it on the conveyor. For all the switches that require no chrome ring or light shield, the operator reaches for the button only when the socket sub-assembly has been placed in cavity (2).

Assembly of the three gang switches is somewhat more complex due to the greater number of parts, however the nature of the operations is similar.

The end of line testing is performed via a fully-automated four-station indexing table. The four stations are (i) a loading station which loads the assembled switch from the conveyor on to the station using a pneumatic pick and place device; (ii) a testing station where the switch is subjected to electrical and force testing, and (for the three gang switches) a barcode label is read; (iii) a camera and laser station where LED illumination intensity and graphic orientation is inspected, and where the customer part number and date code are engraved by laser; and (iv) an unloading station that transfers the switch, using a pick and place device, to a separate conveyor for final inspection, or onto a reject bin.

The final inspection and packaging workstation is fully manual. Here the operator ensures that no scratches, dents or other defects are present on the button's surfaces; checks the integrity of the clipping features; verifies that the terminals are not bent; and ensures that the correct laser marking and date code have been used. Conforming switches are subsequently packed in the respective packaging, whilst non-conforming switches are disposed of in the reject bin.

4.3 Capacity analysis for the current set-up

A capacity analysis was performed in order to quantify the number of switches that can be assembled and tested using the existing production lines. This was done by measuring the time required for every assembly process step of each switch variant, and by analyzing the cycle times and projected production volumes for each individual variant for the four years under consideration. In this respect, the *cycle time* is defined as the time interval for the completion of one complete production unit. In the case study considered, the cycle time was taken to be the longest time from among the three operations performed, i.e. *assembly cycle time* (the time taken to assemble one full switch), *testing cycle time* (the time taken to complete the longest testing step adding the indexing time of the table), and *finishing cycle time* (the time taken to inspect and pack). Equipment availability was assumed to run at 85%, which is due to (i) one product changeover of 15 minutes per shift, resulting in a 3.33% loss; (ii) an allowance of 30 minutes per shift for maintenance activities, including 15 minutes for breakdowns and 15 minutes for planned preventive maintenance, resulting in a 6.67% loss; (iii) personnel related stoppages of 10 minutes per shift resulting in a 2.22% loss; and (iv) process yield running at 97.5%. The number of shifts required to cater for these volumes could thus be calculated using an 85% equipment availability, with 7.5 operating hours per shift, for five days a week and 48 weeks per year.

The results of the capacity analysis are summarized in Table 3. Due to current layout constraints only two product families can be tested in parallel, and this means that the permissible total number of daily shifts is six. As can be seen in the table, during fiscal years 3 and 4, the total output cannot be reached because the number of daily shifts required is not achievable. In addition to this, during fiscal year 4 the number of daily shifts required to achieve the required new three gang switch volumes is 3.11 which is not achievable, since new three gang switches can only be assembled and tested on cell 2. These results pointed out the need of improving the current layouts so as to cater for the required volumes.

Fiscal Year	Number of Daily Shifts Required			Total Number of Daily Shifts Required
	Single Gang	Old Three Gang	New Three Gang	
Year 1	0.30	2.55	0.00	2.85
Year 2	2.28	1.97	0.32	4.57
Year 3	3.38	1.97	1.81	7.16
Year 4	3.75	1.97	3.11	8.83

Table 3. Number of shifts required to cater for the projected volumes using the existing production lines

4.4 Product design specification chart

A PDS chart contains a detailed list of *requirements* that the final product must fulfil, and is drawn up prior to starting the actual design. The aim of the PDS is to encompass all of the required information for a successful solution design and to ensure that the needs of the user are achieved. The PDS also lists a number of *wishes*, which are specifications that are not essential for the success of the project. These wishes however give the project a competitive edge and increase the potential benefits gained through its implementation.

A PDS chart was created for this project, listing all of the specifications that should be taken into account, when designing the required improvements on the switch manufacturing cells. The section of the chart dealing with the *performance* criterion of the production equipment is shown in Table 4. The other criteria that were considered were *target product cost, required service life, serviceability, safety, environment, size, ergonomics, materials, transportation, manufacturing facilities, appearance, quality and reliability, personnel requirements, product lifespan, documentation,* and *commissioning.*

Examples of design wishes (not shown in Table 4) include the minimization of shop floor space occupied by the equipment, the use of inexpensive (but reliable) materials, and the ability to manufacture the equipment in house.

Specification	Requirement	Need
Performance criterion		
Ability to assemble, test and finish the three switch families.	✓	
Ability to cater for the projected volumes, with an excess capacity of 15%.	✓	
An assembly cycle time for the new three gang switches of less than 10 seconds.	✓	
An assembly cycle time for the single gang switches of less than 8 seconds.	✓	
Ability to test single gang and three gang switches simultaneously.	✓	

Table 4. A section of the PDS chart

4.5 Group technology analysis

In this case study, the preferred criterion for parts classification was found to be that based on product characteristics, since the parts fell into three natural groupings as discussed in section 4.1. The characteristics and variations of each of the three part families were analyzed in detail, with a view to confirming this classification and to prepare for the detailed technical design phase of the project.

An analysis of the constituent parts of the single gang switches produced the following results:

Button – there are different types of button, due to different customer requirements, mainly in terms of graphic design, shape and the type of surface. However all the buttons in the switch family have common guiding and clipping features.

Socket – two types of sockets exist with the main geometric difference being the position of the foolproof feature as shown in Figure 7. A second difference is in the socket colour, where type A is black and type B is grey.

PCB – there are different types of PCB having different profiles and different location of the electrical components.

Chrome Ring – there is only one type of chrome ring.

Jewel – the jewel needs to be aligned with the surface so it must have the same shape as the surface of the button. Two types of jewel exist that correspond to two types of surface.

Light Shield – there is only one type of light shield.

Similar analyses were carried out for the old three gang and new three gang switches. The part variations associated with all three product families are summarized in Table 5.

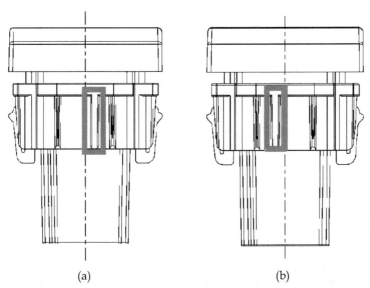

(a)   (b)

Fig. 7. The two types of socket for the single gang switch. (a) Type A, (b) Type B

Single gang switches		Old three gang switches		New three gang switches	
Constituent part	*No. of Variants*	*Constituent part*	*No. of Variants*	*Constituent part*	*No. of Variants*
Button	15	Button set	10	Button set	8
Socket	2	Socket	5	Socket	2
PCB	12	PCB	9	PCB	5
Chrome ring	1	Housing	1	Housing	1
Jewel	2	Slider	1	Metal clip set	1
Light shield	1				

Table 5. Part variations for the three product families

From the results, it can be seen that there are substantial differences between the three sets of products, in terms of the gross geometries and of the constituent parts. In particular, it is noted that a different cavity is required for each of the three families (the geometric differences between the different types of socket within each switch family are minor, and in each case can be catered for by the same cavity). At the same time, the components that constitute each product family allow for ease of automation, since there are only a small number of variations for the parts. The analysis therefore confirmed the classification of the switches into three distinct product families as indicated in Table 5.

4.6 Precedence diagrams

The generalized manufacturing process flow chart for each switch, as extracted from the description given in section 4.2, is illustrated in Figure 8(a), and consists of three major steps. Step 1 involves the assembly of the switch. Step 2 involves the testing of the switch (force, electrical, and illumination testing) and laser marking. Step 3 involves a visual inspection of the switch and final packaging.

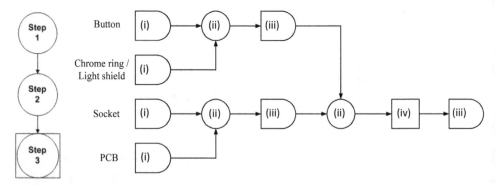

Fig. 8. (a) General process flow chart for each switch, (b) Precedence diagram for the assembly of a single gang switch

The precedence diagram for the assembly process (i.e. for Step 1) for the single gang switch is shown in Figure 8(b). In the figure, the shapes labelled (i) represent temporary storage, the circles labelled (ii) represent the complex operation "bring sub-components together, align, and clip", the shapes labelled (iii) represent delays, and the square labelled (iv) represents a brief visual verification that the clipping has been carried out correctly. For this particular case study it was not possible to simplify these steps any further.

The precedence diagrams for the old three gang and the new three gang switches were extracted in a similar manner. The three diagrams served to guide the development of the morphological chart, and the generation of alternative conceptual solutions, described in the next two sections.

4.7 Morphological chart

A morphological chart is an analytical tool which aims at finding all theoretically conceivable solutions to a problem (Roozenburg & Eekels, 1995). It provides a visual way of capturing the required product functions and of exploring possible different solutions that may exist for each product function. The chart facilitates the presentation of these solutions and provides a framework for considering alternative combinations of the individual function solutions.

The main functions associated with the problem at hand were identified to be the (i) transfer system between stations, (ii) part orienting mechanism, (iii) part feeding mechanism, (iv) handling mechanism, (v) gripping mechanism, (vi) part inspection system, and (vii) part packaging. The morphological chart is shown in Table 6.

Function	Option 1	Option 2	Option 3	Option 4
Transfer System	In-line indexing system	In-Line indexing system with return carriers in the vertical plane	Rotary indexing system	Pallet system
Part Orienting	Vibratory bowl feeder	Magnetic rotary feeder	Machine vision system coupled with a robotic arm	Manual
Part Feeding	Vibrating conveyor	Linear feeder	Horizontal belt conveyor with passive guides	Manual
Handling System	Pneumatic pick and place	Electric pick and place	Robotic arm	Manual
Part Gripping	Vacuum suction	Magnetic gripping	Pneumatic grippers: radial, 3-point and angular	Manual
Part Inspection Systems	Machine vision system	Colour sensor	Human visual inspection	
Packaging System	Robot based system	Customized automation	Manual	

Table 6. Morphological chart for the new manufacturing system. The selected solutions are highlighted.

Solution selection was made on the following bases:

Transfer system – The projected high production volumes necessitate a low indexing time, and if a pallet system is used this would only be achievable by using a very large number of pallets. Thus the manufacturing cost of the system would increase due to the large number of cavities required. A rotary indexing table reduces maintenance interventions, since maintenance requirements are less compared to that of an in-line indexing system.

Part orienting – A vibratory bowl feeder can provide the required output, and is the cheapest and most reliable solution among the four options considered.

Part feeding – High part feeding accuracy is required in order to ensure correct operation of the system and this accuracy can be achieved through the use of linear feeders coupled with vibratory bowl feeders. The use of vibration conveyors or of passive guides cannot achieve the required accuracy. The manual option is expensive.

Handling system – The transportation of the part between two fixed positions can easily be achieved by a pneumatic handling system, which is the cheapest alternative among the four considered.

Part gripping – A system based on vacuum suction could be used, however for its full implementation a number of intricate vacuum heads would need to be designed, thus substantially increasing the cost of the system. A magnetic system cannot be used for all components, since most of the components are made of plastic. A manual system is an expensive option and therefore pneumatic grippers with specifically designed jaws were selected.

Part inspection system – Visual inspection is required at the end of line testing stage (which is already automated, and uses a machine vision system) and at the final inspection station. Due to the complex nature of the final inspection it was determined that this could only be carried out reliably by human operators.

Packaging system – Since the final inspection is manual, the preferred option would be to have the human operator package the completed switch after inspection.

4.8 Concept generation

4.8.1 Overview of proposed concepts

Four different concept layouts were generated to address the problem. The first concept involves automation of the single gang switch assembly, and relocation of all assembly of this switch to Cell 1. Assembly of the new three gang switch would also be automated, and Cell 2 would be dedicated to this process. The second concept involves the retention of the present, labour intensive, assembly processes, but with the incorporation of an additional station to Cell 2 for the assembly of new three gang switches, to meet the projected production volumes. The third concept is a compromise between the approaches of the first and second concepts, and involves the retention of the present, labour intensive, assembly processes for the single gang switch and for the old three gang switch, and the transfer of the single gang switch assembly station from Cell 2 to Cell 1. Cell 2 would be dedicated to the automated production of the new three gang switches as in Concept 1. The fourth concept involves the combination of all production processes into a single cell, and automating the assembly of the single gang and of the new three gang switches. It is noted that due to the fact that the production volume of the old three gang switches is expected to decrease, automation of the assembly process for these switches is not recommended under any of the proposed concepts. The four concepts are presented in greater detail in the following sections.

4.8.2 Concept 1

The proposed layout for Cell 1 under this approach is shown in Figure 9. The cell consists of an indexing table used for the assembly of the single gang switches, two manual jigs used for the assembly of the old three gang switches and a testing indexing table which can test the two different switches simultaneously. Linear conveyors transfer the switches from the assembly stations to the testing station. An operator loads the PCB and button on Station 1 of the loading indexing table. The work carrier of this indexing table will be sub-divided into two sections, one holding the PCB and one the button. These two parts are bought-in parts presented to assembly in painting jigs or trays and therefore automation of the loading function would require a tray changing mechanism and an x-y-z pick and place device. The

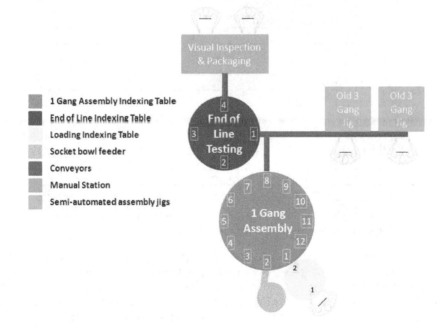

Fig. 9. Proposed Layout for Cell 1 (Concept 1)

initial cost required to create these subsystems would be much higher than the operational cost of one operator who would still be required to attend the machine, and their implementation is therefore not recommended. The parts are then automatically transferred onto a twelve station indexing table. The work carrier on this indexing table will be sub-divided into three sections, namely cavities 1, 2 and 3. The proposed stations for the indexing table are listed in Table 7. The twelve station indexing table was chosen because one is already available at the company, thus reducing the initial cost required. This results in five free stations which can be utilized for future improvements of the layout. The fully assembled single gang and old three gang switches are unloaded onto conveyors which transfer them to the end of line indexing table, based on the current automated system. Table 8 lists the four stations of the fully automated indexing table. The loading station will either pick up one switch type or both switch types, depending upon the switch being available at the conveyor, since old three gang and single gang assemblies would not be synchronized. A new anti-mixing part inspection system, based on machine vision, would need to be incorporated into the end of line testing station, to distinguish between the two switch families.

Station	Description
1	Loading of button onto cavity 3 and PCB onto cavity 1
2	Loading of socket onto PCB
3	Clipping of socket to PCB (sub-assembly 1)
4	Turning of socket and placing onto cavity 2
5	Free station
6	Loading of button onto sub-assembly 1
7	Clipping of button with sub-assembly 1
8	Unloading
9	Free station
10	Free station
11	Free station
12	Free station

Table 7. Single gang switch assembly stations (Concept 1, Cell 1)

Station	Description
1	Loading
2	Force and electrical testing
3	Camera test and laser mark
4	Unloading

Table 8. End of line testing stations (all concepts, all cells)

The proposed layout for Cell 2 consists of a semi-automated twelve station indexing table, used for the assembly of new three gang switches, as shown in Figure 10. The work carrier on this indexing table will be sub-divided into three sections, namely cavities 1, 2 and 3. The socket, clips and housing are oriented via vibratory bowl feeders and automatically loaded on the respective stations. The PCB is manually loaded on a conveyor which is then automatically transferred onto the socket in station 2. Table 9 lists the proposed operations to be performed by each station. The PCB and buttons are manually loaded as in Cell 1. In station 3, the sub-assembled components are unloaded onto a conveyor and the same operator loading the PCB, adds a label to the sub-assembly. Subsequently a second operator adds the three buttons in their corresponding position and places the switch onto a third conveyor. The clipping of the buttons is performed via an automatic clipping and pre-actuation station and finally the switch is loaded onto the testing indexing table. The end of line indexing table is similar to the one on Cell 1.

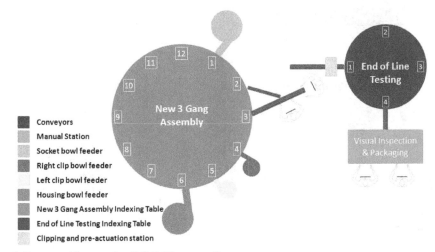

Conveyors
Manual Station
Socket bowl feeder
Right clip bowl feeder
Left clip bowl feeder
Housing bowl feeder
New 3 Gang Assembly Indexing Table
End of Line Testing Indexing Table
Clipping and pre-actuation station

Fig. 10. Proposed layout for Cell 2 (Concept 1)

Station	Description
1	Loading of Socket in cavity 1
2	Loading of PCB onto socket in cavity 1 - (sub-assembly 1)
3	Unloading of sub-assembly 3 onto conveyor from cavity 2
4	Loading of right clip in cavity 3
5	Loading of left clip in cavity 3
6	Loading of housing onto clips in cavity 3
7	Clipping of housing with clips – (sub-assembly 2)
8	Loading of sub-assembly 2 onto sub-assembly 1 in cavity 1
9	Clipping of sub-assembly 2 onto sub-assembly 1 - (sub-assembly 3)
10	Transfer of sub-assembly 3 onto cavity 2
11	Free Station
12	Free Station

Table 9. New three gang switch assembly stations (Concept 1, Cell 2)

4.8.3 Concept 2

The proposed layouts under this approach are shown in Figure 11. This concept entails the incorporation of an additional manual assembly jig to Cell 2 that would be used for the assembly of new three gang switches. This is achieved by modifying the conveyor currently used to transfer the assembled parts from the jigs to the testing indexing table, so as to cater for the additional jig. Changes are also proposed to the testing program of both end of line testers, so as to reduce the testing time required. A new operator is required for the visual inspection and packaging station of Cell 2, so as to cater for all the switches being assembled. Two operators would thus be dedicated to new three gang switches, and one to single gang switches. This concept is a labour intensive concept which however requires less initial investment due to the fact that only minor modifications are required to the existing structure. The projected production volumes can however still be met through this layout.

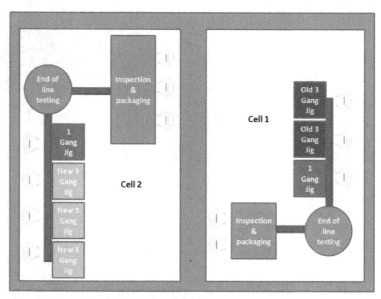

Fig. 11. Proposed layouts for Cell 1 and Cell 2 (Concept 2)

4.8.4 Concept 3

The layouts of the cells under this approach, as described in section 4.8.1, are shown in Figure 12. Modifications are required to the testing program of both cells so as to reduce the testing cycle time.

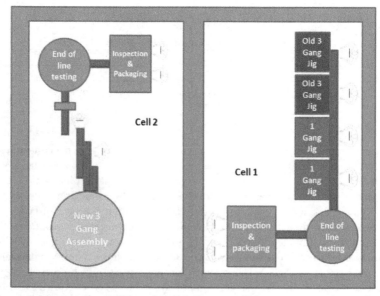

Fig. 12. Proposed layouts for Cell 1 and Cell 2 (Concept 3)

4.8.5 Concept 4

The fourth concept consists of one indexing table that is used for the assembly of both the single gang and the new three gang switches, as shown in Figure 13. A 20-station indexing table is used having work carriers divided into six sections, where three sections (cavities 1, 2 and 3) are used for new three gang switches and the other three (cavities 4, 5 and 6) are used for single gang switches. The main difference between this layout and the one proposed in Concept 1 involves the combination of the two assembly lines. The orienting, feeding, loading and clipping mechanisms are the same as those proposed in Concept 1. The proposed stations are listed in Table 10, where it can be seen that the first twelve stations are dedicated to the new three gang switches and the remaining eight stations to the single gang switches. There would be a total of three free stations. The end of line testing systems are the same as those proposed in Concept 1.

Switch	Station	Description
New Three Gang Switch Assembly	1	Loading of socket in cavity 1
	2	Loading of PCB onto socket in cavity 1 - (sub-assembly 1)
	3	Unloading of sub-assembly 1 onto conveyor
	4	Loading of right clip in cavity 3
	5	Loading of left clip in cavity 3
	6	Loading of housing onto clips in cavity 3
	7	Clipping of housing with clips – (sub-assembly 2)
	8	Loading of sub-assembly 2 onto sub-assembly 1 in cavity 1
	9	Clipping of sub-assembly 2 onto sub-assembly 1 - (sub-assembly 3)
	10	Transfer of sub-assembly 3 onto cavity 2
	11	Free station
	12	Free station
Single Gang Switch Assembly	13	Loading of button onto cavity 6 and PCB onto cavity 4
	14	Loading of socket onto PCB in cavity 4
	15	Clipping of socket to PCB (sub-assembly 4)
	16	Turning of sub-assembly 4 and placing onto cavity 5
	17	Free station
	18	Loading of button onto sub-assembly 4
	19	Clipping of button with sub-assembly 4
	20	Unloading

Table 10. Assembly stations for the single gang and new three gang switches (Concept 4)

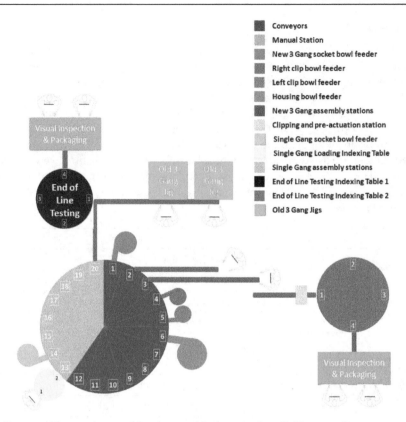

Conveyors
Manual Station
New 3 Gang socket bowl feeder
Right clip bowl feeder
Left clip bowl feeder
Housing bowl feeder
New 3 Gang assembly stations
Clipping and pre-actuation station
Single Gang socket bowl feeder
Single Gang Loading Indexing Table
Single Gang assembly stations
End of Line Testing Indexing Table 1
End of Line Testing Indexing Table 2
Old 3 Gang Jigs

Fig. 13. Proposed layout for combined assembly in a single cell (Concept 4)

4.9 Provisional analytical studies

4.9.1 Overview of analysis

The proposed concepts were analyzed with respect to a number of parameters, namely *cycle time, initial investment cost, labour requirements, line balancing, final product quality, shop floor area consumed, lead time to manufacturing the equipment, maintenance requirements, knowledge transfer availability,* and *flexibility.*

4.9.2 Cycle time

The production of a switch consists of three main operations, namely assembly, testing and finishing. The production cycle time corresponds to the longest cycle time from among these three operations. In order to derive the individual cycle time of each operation, the expected duration of every manufacturing step for each of the proposed concepts was estimated, and the operation cycle time is then given by the longest duration from among its individual stations, taking into account also the indexing time where applicable. It was found that the bottleneck of the production line for all four concepts is the assembly operation. The results of the analysis are summarized in Table 11.

	Single gang	Old three gang	New three gang
Concept 1	6.0s	15.0s	6.0s
Concept 2	8.5s	15.0s	10.8s
Concept 3	8.5s	15.0s	6.0s
Concept 4	6.0s	15.0s	6.0s

Table 11. Production cycle times for each switch family under each concept

4.9.3 Initial investment costs

The initial costs associated with each concept were estimated. It was found that Concept 4 would be the most expensive to implement, followed by Concept 1, Concept 3, and Concept 2 respectively.

4.9.4 Labour requirements

The total number of human operators required for each conceptual approach was determined. Based on the descriptions given in section 4.8, Concept 1 would require nine operators, Concept 2 would require twelve operators, Concept 3 would require ten operators, and Concept 4 would require nine operators.

4.9.5 Line balancing

The *balance efficiency* of a production line is a measure of the time used for productive work at each station as compared to the total available time (e.g. Groover, 2001). In a perfectly balanced line (i.e. 100% balance efficiency) the durations of the jobs carried out at each individual station (be they manual or automated) would be exactly equal to each other (and equal to the cycle time), and there would be no idle time at any station. In practice this ideal situation is very difficult to obtain since it is unlikely that the total work required can be broken down into discrete steps of exactly the same duration, however it remains important to strive for as high an efficiency as possible. The achievable line balance efficiencies for the four concepts described in section 4.8 are given in Table 12.

	Single gang	New three gang
Concept 1	63%	68%
Concept 2	60%	59%
Concept 3	60%	68%
Concept 4	68%	

Table 12. Balance efficiencies for the single gang and new three gang switch families under each concept

4.9.6 Final product quality

The increase in production rate should not be achieved at the expense of reduced product quality and therefore all concepts considered were developed with great concern towards maintaining high quality standards. Thus for example, the proposed automatic handling of buttons would be performed without any part coming into contact with the surface of the button. Given that a well designed automation system is often capable of greater consistency than a system based on human operators, it would be expected that product quality would increase through greater use of automation.

4.9.7 Shop floor area consumed

Due to limitations in the shop floor area, the space consumed by the manufacturing lines would need to be minimized. It was estimated that approximately 55 m² of floor area would be consumed by the cells under Concept 1, 50 m² under Concept 2, 50 m² under Concept 3, and 90 m² under Concept 4.

4.9.8 Lead time to manufacturing the equipment

The development of a new production line involves the mechanical and electrical systems design, manufacture of the required components, wiring and programming of the system, and assembly, tuning, and testing of the system. It was estimated that for Concept 1 the total lead time would be approximately 28 weeks, for Concept 2 it would be 2.5 weeks, for Concept 3 it would be 23 weeks, and for Concept 4 it would be 29 weeks.

4.9.9 Maintenance requirements

The selected system would require both preventive and corrective maintenance tasks in order to function correctly over a period of time. The time required to perform such maintenance translates into lost production time and therefore the maintenance requirements of the developed concept need to be minimized so as to maximise productivity and efficiency. The maintenance requirements increase with the number of mechanical, pneumatic and electrical components in the system. Thus Concept 1 and Concept 4, which contain indexing tables, automated stations and vibratory bowl feeders would have higher maintenance requirements than would Concept 2 which contains only mechanized jigs.

4.9.10 Knowledge transfer availability

In today's competitive market, cost reduction and shorter lead time to market are very important. Knowledge transfer aims at achieving these goals, through sharing of the knowledge learnt from previous projects, especially in terms of technologies and procedures. The four conceptual solutions are based on layout styles that are already widely applied at the company and therefore personnel are already experienced with similar equipment. This results in a reduction of the lead time to implement the concept and an improvement in operation and troubleshooting efficiency.

4.9.11 Flexibility

The number of distinct members of the three switch families is expected to increase in the next few years. All four concepts that have been generated allow for these expected new variations, however future customer requirements are difficult to forecast with precision. The *automation* concepts that have been proposed are based on fixed automation systems and thus would not allow for major variations in switch design. An increased flexibility is however supplied by the *manual* assembly jigs since in the associated concepts, potential future changes to the switch design can be more easily catered for by the human operators.

4.10 The decision matrix

Concept selection was based on a decision matrix. The ten criteria discussed in section 4.9 were ranked in order of importance, in consultation with experienced company personnel, and were subsequently given a weighting ranging from 10 for the most important criterion, to 1 for the least important. Each concept was then assigned an individual score for each criterion, based on the analysis of section 4.9. The total score for each concept was then obtained from the weighted sum of the individual scores. The complete decision matrix is shown in Table 13. Based on this result, the selected solution was based on Concept 1.

Selection criterion	Weighting	Concept 1	Concept 2	Concept 3	Concept 4
Labour requirements	10	7	1	3	7
Cycle time	9	7	1	3	7
Final product quality	8	7	1	3	7
Initial investment cost	7	3	7	5	1
Line balancing	6	5	1	3	7
Knowledge transfer availability	5	3	3	3	3
Flexibility	4	1	5	3	1
Lead time to manufacture	3	3	7	5	1
Shop floor area consumed	2	5	7	7	1
Maintenance requirements	1	3	7	5	3
TOTAL		281	159	195	265

Table 13. The decision matrix

4.11 Process failure modes and effects analysis

A process failure modes and effects analysis (PFMEA) is a detailed analysis of the errors and malfunctions that can occur during an engineering process, including assessment of the severity, probability of occurrence, and effects of the potential malfunctions, with a view to improve the process design and reliability. An extensive PFMEA was carried out on the selected concept, searching for and assessing various potential failure modes at every station of both production cells. In addition to the various specific process design provisions that were made to address each failure mode that was identified and evaluated through the PFMEA, a number of general conclusions could be drawn from the qualitative and quantitative results of the exercise. Firstly, it was noted that various mistakes can be made by the human operators at the manual stations, and that these mistakes can be minimized by providing clear and concise working instructions to the operators. In this respect all necessary training must also be given. Secondly, it was noted that malfunction of the grippers and air supply, and errors in alignment and settings, can have significant but avoidable detrimental effects on the production process. During the PFMEA a high severity rating was assigned to all of the pick and place operations, to motivate special attention to all associated production line components during commissioning. These ratings would later need to be revised so as to reflect better the final conditions of the line. Thirdly, due to the fact that numerous variants exist for the parts being assembled, the risk of product misidentification and mixing is high. Therefore inspection tests need to be performed in order to detect this failure mode, and this can be achieved through the addition of colour sensors on the linear feeder, a camera inspection on the end of line testing, and an automatic bar code scanner. Fourthly, it is noted that as a final measure, all potential failure modes can be detected by the end of line tests being performed, thus ensuring high reliability of the final product being delivered to the customer.

4.12 Safety analysis

In order to ensure that all safety considerations are integrated within the project as early as possible in the design process, a safety analysis was performed on the selected concept. The analysis followed the five step approach recommended by Bahr (1997) – Step 1: Define the system; Step 2: Identify the hazards; Step 3: Evaluate the hazards; Step 4: Resolve the hazards; and Step 5: Carry out follow-up activity. The system was defined (Step 1) to encompass the two production lines (Cell 1 and Cell 2). The results of Step 2 through Step 4 of the safety analysis are summarized in Table 14. Step 5 can be realized through continued regular checks to ensure: effectiveness of all safety modules; correct functionality of all emergency stop buttons; presence of all protective covers and that all covers are tightly fixed; no cutting edges have been created by wear and tear of the machine; presence of all required grounding systems; and effectiveness of the extraction system and regular filter replacement.

4.13 Ergonomic analysis

In order to improve worker interaction with the system being operated, ergonomic principles were applied to the system design, so as to accommodate human needs. This improves operator performance and well-being, resulting in an increase in overall system performance and efficiency. The analysis has been performed on the manual workstations to ensure that the most ergonomic design is chosen. The ergonomic design specifications are based on the recommendations in Kanawaty (1992) and Wojcikiewicz (2003).

The height of the seated workbench is to be set at approximately 0.72 m so as to ensure that the worker's arms are below the shoulders. The leg clearance should be approximately 0.4 m at knee level and 0.6 m for the feet, without any obstructions such as drawers, between the legs. Height adjustable chairs are to be utilised for the all manned workstations, so as to ensure that the back and neck are not inclined more than 30°. A foot rest should also be available for operators if required. All silos and trays containing assembly parts should be placed within the maximum reach area of the operator, whereas the cavity should be placed within the optimum reach area. Movement of the eyes should be minimized since it takes approximately three seconds for the eyes to rotate and re-focus. Therefore the buttons and the work-piece should be placed within the 15° view angle, on either side of the centreline, since this angle requires no eye movement to allow for the grabbing of the parts. Part silos and the label printer should be placed within the 35° view angle. Correct lighting should be available since this helps to reduce errors and thus improve productivity. The light intensity requirement for the operations to be performed in this case study should be about 500 lux, where one lux is given by the illumination of a surface placed one meter away from a single candle. The light should be uniformly distributed so as to avoid pronounced shadows and excessive contrasts.

Category	Hazard description	Potential causal factors	Sev.	Occ.	Hazard Resolution
Mechanical	Crushing of body part	Unexpected movement of pneumatic cylinders	II	A	Safety guards with interlocks
		Unexpected movement of electric motors on testing station	II	A	
	Operator cuts a body part	Sharp Edges on equipment	III	B	Chamfers and edge deburring
	Operator catches a body part in a pinch point	Pulleys controlling conveyor movement	II	B	Protective covers for all pulleys
	Impact	Unexpected movement of pneumatic cylinders	II	B	Safety guards with interlocks
		Unexpected movement of indexing table	II	B	
	Wrap Points	Entanglement of clothing and accessories with conveyor	III	C	Protective covers for all pulleys; emergency stops
Electrical / Electronic	Energized equipment resulting in electric shock	Improper electrical connections and wiring	I	C	Include fuses, circuit breakers, and electrical grounding; use electrical safety checklist with double-checking; enclose wiring in control box; emergency stop switches
		Poor insulation	I	D	
		Insufficient grounding	I	D	
		Inadvertent activation	I	B	

Severity key: I-Catastrophic; II-Critical; III-Marginal; IV-Negligible.
Occurrence key: A-Frequent; B-Probable; C-Occasional; D-Remote; E-Improbable

Table 14. (first part) Safety analysis: identification, evaluation, and resolution of hazard

Category	Hazard description	Potential causal factors	Sev.	Occ.	Hazard Resolution
Noise / Vibration	Permanent damage to hearing	Environmental sound level exceeds 80dBA	II	B	Pneumatic cylinders equipped with silencers.
	Personnel fatigue	Excessive vibrations to operator's workstation	III	A	No vibratory or linear feeders placed in proximity to operators' workstations.
Lasers	Eye exposure	Collimated beam direct from the laser head into the operator's eyes	II	C	Laser systems enclosed by safety guards with interlocks
	Burning of operator hands	Collimated beam direct from the laser head over the operator's hand	II	C	
	Operator inhales toxic fumes	Toxic fumes arising from burning of plastics by laser marking.	III	A	Fume extraction system

Severity key: I-Catastrophic; II-Critical; III-Marginal; IV-Negligible.
Occurrence key: A-Frequent; B-Probable; C-Occasional; D-Remote; E-Improbable

Table 14. (continued) Safety analysis: identification, evaluation, and resolution of hazards

4.14 New capacity analysis

A detailed capacity analysis was carried out on the proposed production system, based on the assumptions made in section 4.3. The results of this analysis are summarized in Table 15. The total required output can be reached easily using the proposed system, and even in the most demanding year (Year 4) there is a substantial reserve capacity.

	Number of Daily Shifts Required			Total Number of Daily Shifts Required
Fiscal Year	Single Gang	Old Three Gang	New Three Gang	
Year 1	0.15	2.55	0.00	2.70
Year 2	1.02	1.97	0.16	3.15
Year 3	1.54	1.97	0.94	4.45
Year 4	1.67	1.97	1.63	5.27

Table 15. Number of shifts required to cater for the projected volumes using the proposed production lines

4.15 Provisional return on investment analysis

In order to estimate the financial benefits that would be gained by the company upon the implementation of the proposed layouts, a provisional return on investment analysis was carried out. The operational cost savings were calculated by comparing labour costs under the present and the proposed layouts. The labour seconds required to manufacture each switch was calculated by multiplying the cycle time by the number of operators required to operate the cell. These calculations indicate substantial cost savings over the four year period, with return on the initial investment achieved in less than three years.

5. Conceptual drawings

While not included among the more critical procedural guidelines proposed in section 3 above, the generation of three-dimensional renditions of the conceptual design of a system helps the design team visualize the overall concept and may aid in the optimization of the spatial layout.

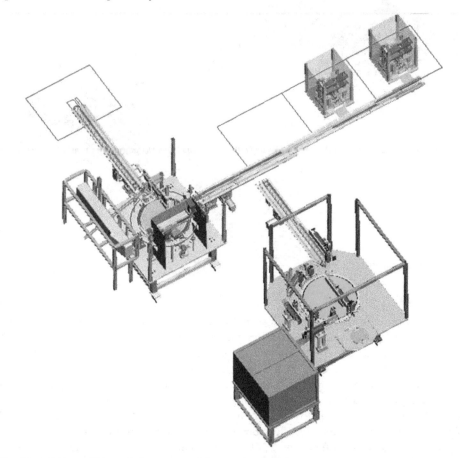

Fig. 14. A 3-D rendition of the proposed layout for Cell 1

A three-dimensional conceptual drawing of the proposed Cell 1 for this case study is given in Figure 14, and shows the two mechanized old three gang switch assembly jigs, the indexing table for single gang switch assembly, and the end of line testing module. The drawing for the proposed Cell 2 is given in Figure 15, and shows the indexing module for the assembly of the new three gang switches and the end of line testing module. These drawings were generated using Pro/ENGINEER (PTC, 2008).

Fig. 15. A 3-D rendition of the proposed layout for Cell 2

6. Conclusion

The procedural guidelines that have been presented in this work contribute an important planning and implementation approach for the development of a conceptual design for a new manufacturing system, when migrating from manual to automated assembly of a part family of products. The novelty in the approach presented here is in the fusion of the conventional guidelines for the development of production automation systems, with a product design approach to the manufacturing system. The detailed case study that is presented in this work serves to demonstrate the application of the guidelines, and will serve as a useful reference tool for future projects of this nature. Future research in this area can include an extension of this approach to the embodiment and detailed design stages of the production system development.

In the case study it is shown that the new automated manufacturing system will result in a cycle time reduction of six seconds for the single gang switches, and of nine seconds for the new three gang switches. This will result in a corresponding increase in production capacity, thus also improving the flexibility of the company since it will be able to react to new customer orders more quickly. The new layout also results in a reduction in the manufacturing lead time, allowing the forecasted customer requirements to be catered for with over 40% excess capacity. The initial investment that is required is justified, since a significant reduction in labour costs is experienced, resulting in a return on investment of less than three years.

7. References

Andreasen, M.M. & Hein, L. (1987). *Integrated Product Development*, IFS (Publications), ISBN 0948507217, Bedford, UK.

Asfahl, C.R. (1992). *Robots and Manufacturing Automation* (Second Edition), John Wiley & Sons, Inc., ISBN 0471553913, NY, USA.

Bahr, N.J. (1997) *System Safety Engineering and Risk Assessment: A Practical Approach*, CRC Press, ISBN 1560324163, London, UK.

Baines, T. (2004). An integrated process for forming manufacturing technology acquisition decisions. *International Journal of Operations & Production Management*, Vol. 24, No. 5, pp. 447–467, ISSN 0144-3577.

Boothroyd, G., Dewhurst, P. & Knight, W. (2001). *Product Design for Manufacture and Assembly* (Second Edition, Revised and Expanded), CRC Press, ISBN 082470584X, NY, USA.

Chan, F.T.S & Abhary, K. (1996). Design and evaluation of automated cellular manufacturing systems with simulation modelling and AHP approach: a case study. *Integrated Manufacturing Systems* Vol. 7, No. 6, pp. 39–52, ISSN 0957-6061.

Dieter, G.E. & Schmidt, L.C. (2009). *Engineering Design* (Fourth Edition), McGraw-Hill, ISBN 0072837039, NY, USA.

Groover, M.P. (2001). *Automation, Production Systems, and Computer-Integrated Manufacturing* (Second Edition), Prentice-Hall, ISBN 0130889784, NJ, USA.

Hyer, N. & Wemmerlov, U. (2002). *Reorganizing the Factory: Competing through Cellular Manufacturing*. Productivity Press, ISBN 1563272288, OR, USA.

Kanawaty, G. (ed.) (1992). *Introduction to Work Study* (Fourth Edition), International Labour Organization, ISBN 9221071081, Geneva, Switzerland.

Kapp, K.M. (1997). The USA Principle: The Key to ERP Implementation Success. *APICS~The Performance Advantage*, Vol. 12, pp. 62–66.

PTC (2008). *Pro/ENGINEER design software*, Parametric Technology Corporation, MA, USA.

Roozenburg, N.F.M. & Eekels, J. (1995). *Product Design: Fundamentals and Methods*, John Wiley & Sons, Inc., ISBN 0471954659, NY, USA.

Säfsten, K., Winroth, M. & Stahre, J. (2007). The content and process of automation strategies. *International Journal of Production Economics*, Vol. 110, pp. 25–38, ISSN 0925-5273.

Ullman, D.G. (1997). *The Mechanical Design Process*. McGraw-Hill, ISBN 0070657564, NY, USA.

Winroth, M. & Säfsten, K. (2008). Automation Strategies - Implications on strategy process from refinement of manufacturing strategy content. *Proceedings of the 19th Annual Conference of the Production and Operations Management Society*, 008-0708, La Jolla, CA, USA, May 2008.

Wojcikiewicz, K. (2003). Seven Key Factors for Ergonomic Workstation Design. *Manufacturing Engineering*, Vol.131, No.1, pp. 45–50.

7

Stochastic Capacitated Cellular Manufacturing System Design with Hybrid Similarity Coefficient

Gökhan Eğilmez[1], Gürsel A. Süer[1,*] and Orhan Özgüner[2]
[1]*Ohio University*
[2]*Johns Hopkins University*
USA

1. Introduction

Manufacturing system design is one of the most crucial steps of business processes. Several approaches have been proposed and implemented to increase productivity and profitability due to the change in customer characteristics, market condition and economy. Cellular Manufacturing (CM) is one of the approaches that emerged as an application of Group Technology in the late 70s due to the increase in product variety and demand variance. Group Technology (GT) is a product-oriented manufacturing approach to group similar products for smaller batch size production. As an application of GT, CM is the physical or virtual division of manufacturing facilities into manufacturing cells. A manufacturing cell is a small group of machines and/or workers ideally arranged in a flow layout to produce "similar items", in other words "product families".

Production volume and product variety have significant impact on the design of manufacturing system. Layout of the shop floor is generally used to classify the manufacturing systems. There are four well-known layout types; namely: fixed layout, product layout, process layout and cellular layout. A fixed layout consists of fixed parts and non-fixed resources which travel to parts to perform the operations. In product layout, resources are arranged based on the sequence of operations. This layout is very efficient to meet high volume demand when product variety is low. On the other hand, in process layout, similar resources thus processes are grouped together to meet low and medium volume demand and high product variety. Product layout is more efficient in terms of material flow, whereas process layout is more flexible to deal with high product variety. Cellular layout is a hybrid layout which includes the advantages of both product and process layouts. Cellular layout improves the manufacturing system performance from many aspects such as reduction in material handling, lead times, work-in-process inventory (WIP), re-work, scrap and efficient floor space usage (Wemmerlov U. & Johnson D. J., 1997).

Even though most of the cellular manufacturing system (CMS) design approaches work with deterministic data, uncertainty indeed significantly influences the CMS performance,

---

* Corresponding Author

especially in labor-intensive manufacturing cells. Therefore, the impact of such design parameters as variance in demand and variance in processing times should be taken into consideration during the CMS design. Moreover, most of the works in literature deals with the cell and family formation problem from only route-based similarity point of view. However, since the demand variance can have ruining impact on system performance, demand-based similarity should be also taken into consideration when building similarity matrix. In this chapter, a hybrid similarity matrix, which incorporates route-based and demand-based similarities, is proposed and a stochastic non-linear mathematical model is developed to design CMS considering uncertain demand and processing times. To validate the proposed model, simulation experiments are carried out. Finally, a Genetic Algorithm approach is proposed to deal with large problems.

2. Literature review

The literature is abundant with the works that include optimization methods. In addition to mathematical models, heuristics and meta-heuristics are used to tackle larger problems. The majority of works in literature address deterministic CMS design. However, uncertainty in some parameters such as demand and processing times brings probabilistic nature to design problems. While most of the studies in the literature have addressed the deterministic CMS design problem, less attention is paid to the problems that consider the probabilistic demand and processing times. The literature is reviewed in two sections, namely: deterministic design and stochastic design.

2.1 Deterministic CMS design

In deterministic case, mathematical optimization techniques are used to solve the cell formation problem. As a preliminary work, Purcheck (1974) developed a mathematical classification for the systematic analysis of process routes to group technology and cell formation problem (Purcheck, 1974). Kusiak (1987) provided a comparison of matrix and integer programming models, and discussed the impact of the models on the quality of process families and machine cells (Kusiak, 1987). Shtup (1989) proved the equivalency of cell formation problem to the Generalized Assignment Problem (GAP) (Shtubt, 1989). Rajamani, Singh and Aneja (1990) studied the impact of alternative process plans on the resource utilization and developed three integer programming models to analyze the effect of alternative process plans and simultaneous formation of part families and machine groups (Rajamani, Singh, & Aneja, 1990). Wei and Gaither (1990) developed an integer programming model for cell formation problem (Wei & Gaither, 1990). The objective was to minimize the cost of manufacturing exceptional parts outside the cellular system, subject to machine capacity constraints. Shafer and Rogers (1991) proposed a goal programming model to CMS design problem with the objectives: reducing setup times, minimizing intercellular movements of products and the investment in new equipment, and maintaining acceptable machine utilization levels (Shafer & Rogers, 1991). Kamrani, Parsaei and Leep (1995) developed a mathematical model and tested the performance of the model with simulation in four phases, namely: coding of parts, family formation, resource optimization, simulation (Kamrani, Parsaei, & Leep, 1995). Heragu and Chen (1998) applied a mathematical model to cell formation problem by considering three aspects; resource utilization, alternate routings, and practical constraints (Heragu & J. Chen, 1998). Chen (1998) worked on designing a sustainable cellular manufacturing system in a dynamic environment and developed an integer programming model to minimize material handling and machine costs as well as cell reconfiguration cost for a multi-period planning horizon (Mingyuan Chen, 1998). Wang (1998) formulated a linear assignment model to the group formation problem (Wang, 1998). Sofianopolou (1999) proposed a mathematical model and a two-phased simulated annealing algorithm to solve the problem of grouping machines into cells and selecting a unique product process plan for each product to be produced (Sofianopoulou, 1999). The manufacturing systems considered include such features as replicate machines and several design requirements as well as operation sequence constraints. Akturk and Turkcan (2000) developed an integrated algorithm that considers the cell layout, part-family and cell formation problems simultaneously (Akturk M.S. & Turkcan A., 2000). Albadawi, Bashir and Mingyuan (2005) proposed a two-phased mathematical model for cell the formation problem (Albadawi, Bashir, & Mingyuan Chen, 2005). In the first phase, factor analysis is used to build similarity matrix and machine cells are identified. In the second phase, parts are assigned to the identified machine cells with an integer programming model.

Metaheuristics have also been used to deal with larger cell formation problems. The most commonly used ones are Genetic Algorithms (GA), Simulated Annealing (SA) and Tabu Search (TS). Genetic Algorithms is a random search technique which generates solutions by using techniques inspired by natural evolution. Simulated Annealing is another search based optimization technique which evolves by replacing the current solution by a random "nearby" solution to reach a near global optimum. Tabu Search is a local neighborhood search technique which improves the solution quality by modifying the neighborhood structure of each solution as the search progresses. Moon, Gen and Süer (1999) developed a GA model to minimize additional capital investment in manufacturing cell design (Moon, Gen, & Suer, 1999). Asokan, Prabhakaran and Kumar (2001) proposed two metaheuristics, GA and SA, for the cell formation problem with the objective of minimizing the total moves and minimizing the cell load variation (Asokan, Prabhakaran, & Satheesh Kumar, 2001). Süer, Pena and Vazquez (2003) developed an evolutionary algorithm and applied to three different problems with seven different cost schemes with the objective of minimizing the total machine investment cost (Suer, Pena, & Vazques, 2003). Cao and Chen (2004) formulated an integrated methodology, which consists of a mixed integer non-linear programming model and a TS algorithm for the NP-Hard Problems (Cao & M. Chen, 2004). Jayaswal and Adil (2004) added simulated annealing and local search heuristics to minimize the sum of costs of inter-cell moves, machine investment and machine operating costs (Jayaswal & Adil, 2004). Solimanpur, Vrat and Shankar (2004) modeled a multi-objective integer programming and GA with multiple fitness functions to the design of cellular manufacturing systems with independent cells (Solimanpur, Vrat, & Shankar, 2004).

2.2 Stochastic CMS design

In contrast to abundant literature on deterministic CMS design, only a handful of works dealt with uncertainty. Such stochastic parameters as demand, processing time, capacity requirements are the driver of uncertainty in manufacturing environment. Seifoddini (1990) dealt with the uncertainty of the product mix and developed a probabilistic model to minimize the expected intercell material handling costs of the system (Seifoddini, 1990). Harhalakis, Nagi and Proth (1998) studied minimizing the expected inter-cell material handling cost over the entire design horizon and developed a two-stage heuristic approach (Harhalakis, Nagi, & Proth, 1998). In the first stage, the production volumes are determined with respect to the joint probabilities for every feasible production mix; in the second stage, the cell formation is obtained via the heuristic method. Wicks and Reasor (1999) employed forecasting methods to determine product mix and the demand for products and solved the multi-period cell formation problem with GA (Wicks & Reasor, 1999). Saad (2003) addressed reconfiguration of manufacturing systems and developed following sub-modules; configuration and reconfiguration module, loading module, and simulation-based scheduling module (Saad, 2003).

Queuing theory is also applied to cell formation problem (Mehrabad and Ghezavati, 2009). Each machine is considered as server and each product is assumed as customer. The objective is to minimize the idleness costs for machines, the total cost of sub-contracting for exceptional elements and the cost of resource underutilization. Süer et al. (2010) proposed both deterministic and stochastic approaches in CMS design. Their stochastic approach considered uncertainty in both product demand and production rates (Suer, Huang, & Sripathi, 2010). In this approach, a layered cellular design concept is introduced to cell formation problem. Cells are identified as dedicated, shared and remainder cell to deal with the uncertainty and a product family can be assigned to more than one Cell. In their study, the generalized p-median model by (Kusiak, 1987) is modified to meet objectives as maximizing the utilization of cells and forming the most similar parts as families. However, cell formation model considers the capacity requirements as deterministic even though there is uncertainty in demand and production rates thus capacity requirements.

In this chapter, a stochastic capacitated p-median model is developed to deal with the probabilistic demand and production rates, thus capacity requirements based on the Süer et al.'s (2010) deterministic approach. A new similarity coefficient is defined to combine the demand and process similarity. A new Genetic Algorithm (GA) model is developed for the larger problems. The obtained cell configurations from stochastic mathematical model and GA are simulated with Arena Simulation Software.

3. The manufacturing system studied

The problem is derived from a jewelry company. There are thirty products and eighteen machines in the system. Each product has to be processed on several machines depending on its process route. Since each product's route represents a unidirectional flow, the cell configuration is flow shop. The machine with the maximum processing time among all machines on process route is the bottleneck machine.

Each cell in the system is allocated to only one product family. In other words, cells are independent and dedicated to one product family. Hence, inter-cell transfer of products is not allowed. Inter-cell transfer restrictions have been also used in several manufacturing systems such as pharmaceutical, medical device and food manufacturing. In some of these industries, independent cell configuration is inevitable since potential product mix up may cause serious problems. Each product can only be assigned to one cell (no product splitting is allowed among cells). Since machine setup times are negligible, they are assumed to be zero in this study. Annual production capacity is taken as 2000 hours (50 weeks/yr * 40 hours/week). The annual demand and processing time for each product are random and follow normal distribution. The problem is the identification of product families and corresponding dedicated cells considering stochastic demand, stochastic processing times and hybrid similarity coefficient.

4. The proposed solution methodology: Stochastic CMS design

The proposed solution methodology is a hierarchical one and it consists of five steps, namely: identification of similarity coefficients, determining the bottleneck machine, and determining the probabilistic capacity requirements, stochastic non-linear mathematical model, and simulation. An example problem is solved to explain the methodology used.

4.1 Identification of similarities

In this section, identification of similarities is explained. Three types of similarity coefficients are used, namely: route-based, demand-based and hybrid similarity. Route- based similarity coefficient only considers the processing similarities of products in the manufacturing system. Demand-based similarity only considers the demand variation among products. Hybrid similarity is the combination of both similarity coefficients. Both of the similarity coefficients are explained in detail in the following sections.

4.1.1 Route-based similarity

The route-based similarity matrix is constructed based on the route similarities among products. Süer et al. (2010) modified the McAuley's (1972) similarity coefficient definition to find the similarities among products. The similarity coefficients are calculated via the suggested equation by Süer et al. (2010) as shown in equation 1. The route-based similarity (RB_{ij}) between products i and j is the ratio of number of common machines to total number of machines required.

$$RB_{ij} = \frac{No.of \ machines \ processing \ both \ parts \ i \ and \ j}{No.of \ machines \ processing \ parts \ either \ i \ or \ j} \tag{1}$$

4.1.2 Demand-based similarity

The main motivation of this similarity measure is to identify stable and unstable products. Stable products have lower variability and unstable products have higher variability. Assigning stable and unstable products to the same cell can cause turbulence in the cell. Even a single unstable product can complicate the operation control issues in a cell. Therefore stable products and unstable products are separated and allocated to different cells. By doing this; the turbulence in CMS is restricted to cells with unstable products only.

In this similarity coefficient, products' similarities are calculated based on the variability in demand (Equation 2). The variability in demand for product i (Vd_i) is obtained dividing mean demand (μd_i) by the variance of demand ($\sigma^2 d_i$) as shown in Equation 2 (Silver & Peterson, 1985). Firstly, the demand variability is calculated for all products.

$$Vd_i = \frac{\mu d_i}{\sigma^2 d_i} \qquad (2)$$

Secondly, the absolute difference between each pair of products' variability values is obtained and entered in the difference matrix. Thirdly, the obtained difference values are scaled from 0 to 1 to be converted to demand-based similarity coefficients of pairs. In other words, the variability difference matrix is converted to variability dissimilarity matrix to be used as demand-based similarity matrix. The maximum difference that a pair has in difference matrix is assumed as the greatest dissimilarity. The scaling is applied assuming the maximum difference as 1. Fourthly, the dissimilarity matrix is converted to similarity matrix by subtracting dissimilarity values from 1.

4.1.3 Hybrid similarity

There is a need to strike a balance between route-based similarity vs demand-based similarity. Hybrid similarity coefficient is developed to cover both of previously explained similarities. Equation 3 represents the calculation of similarity coefficient. Beta (β) and ($1 - \beta$) are the proportional impacts of route-based and demand-based similarities on the hybrid similarity coefficient, respectively. In this study, Hybrid Similarity Coefficient is used in CMS design.

$$H_{ij} = \beta * RB_{ij} + (1 - \beta) * DB_{ij} \qquad (3)$$

4.1.4 Hybrid similarity example

An example is derived to illustrate how the similarity concept is applied. Assume that there are five products with the following route, probabilistic demand and demand variability information shown in Table 1. According to the route information given in Table 1, route-based similarities are calculated by using Equation 1. The route-based similarity matrix is shown in Table 2. Demand-based similarity is the second step to calculate the hybrid similarity.

To build demand-based similarity matrix, first of all the difference between the mean demand/variance of demand ratios (Vd_i) are calculated for all pairs and shown in Table 3. The maximum difference is 4.757 and the minimum difference is 0. These values are scaled to 0-1 range as shown in Table 4. These values are then subtracted from 1 and thus the dissimilarity matrix given in Table 4 is converted to demand-based similarity matrix given in Table 5.

Product	Opr. 1	Opr. 2	Opr. 3	Mean Demand	Variance of Demand	Vd_i
1	A	B	C	2999	1284	2.336
2	A	C		4297	2604	1.650
3	A	C	D	2217	346	6.408
4	B	C	E	1255	359	3.496
5	D	F		2463	454	5.425

Table 1. Operational routes and demand information

Product \ Product	1	2	3	4	5
1	-	0.67	0.5	0.5	0
2	0.67	-	0.67	0.25	0
3	0.5	0.67	-	0.2	0.25
4	0.5	0.25	0.2	-	0
5	0	0	0.25	0	-

Table 2. Route-based similarity matrix

Product \ Product	1	2	3	4	5
1	0.000	0.686	4.072	1.160	3.089
2	0.686	0.000	4.757	1.846	3.775
3	4.072	4.757	0.000	2.912	0.982
4	1.160	1.846	2.912	0.000	1.929
5	3.089	3.775	0.982	1.929	0.000

Table 3. Vd_i Difference Matrix

Scaled Matrix	1	2	3	4	5
1	-	0.144	0.856	0.244	0.649
2	0.144	-	1	0.300	0.793
3	0.856	1	-	0.612	0.207
4	0.244	0.388	0.612	-	0.406
5	0.649	0.793	0.207	0.612	-

Table 4. Scaled Difference Matrix

Product \ Product	1	2	3	4	5
1	-	0.856	0.144	0.756	0.351
2	0.856	-	0.000	0.612	0.207
3	0.144	0	-	0.388	0.793
4	0.756	0.612	0.388	-	0.594
5	0.351	0.207	0.793	0.594	-

Table 5. Demand-based similarity matrix

After both route-based and demand-based similarity matrices are built, hybrid similarity matrix is developed by using Equation 3. In this example, β is taken as 0.5. The developed hybrid similarity matrix is shown in Table 6.

Product \ Product	1	2	3	4	5
1	-	0.763	0.322	0.628	0.175
2	0.763	-	0.335	0.431	0.103
3	0.322	0.335	-	0.294	0.522
4	0.628	0.431	0.294	-	0.297
5	0.175	0.103	0.522	0.297	-

Table 6. Hybrid similarity matrix

4.2 Bottleneck machine identification

In this case, the definition of bottleneck machine is modified since processing times are probabilistic. An example is given in Table 7 to illustrate the situation for product i.

Product i	Opr. - 1 on M/C 1	Opr. - 2 on M/C 2
Mean (μ)	5 min	4 min
Standard Deviation (σ)	1.2	1.6
Process Time Estimate based on 2 Sigma (ε=2)	7.4 min	7.2 min
Process Time Estimate based on 3 Sigma (ε=3)	8.6 min	8.8 min

Table 7. Bottleneck machine identification

Assume that product i requires two operations and the mean processing times for operations 1 and 2 are 5 min and 4 min, respectively. Also assume that standard deviation of processing time for operation 1 is 1.2 and for operation 2 is 1.6 minutes. If only mean values are to be considered, machine 1 would be regarded as the bottleneck machine. If processing times are estimated based on 2 sigma (ε= 2) using Equation (4), then processing time estimates will be 7.4 min and 7.2 min, respectively and machine 1 will still be the bottleneck machine. However, if the processing times are estimated based on 3 sigma (ε= 3), then the bottleneck operation will shift to machine 2. In this paper, we have considered the processing time estimate based on 3-sigma level as the basis for the bottleneck machine identification. The reason for this is that the probability that actual processing time will exceed the estimate based on 3-sigma value is very small.

$$p_{ik}^{e} = \mu_{ik} + \varepsilon * \sigma_{ik} \tag{4}$$

where, p_{ik}^{e} is the processing time estimate, μ_{ik} is the mean processing time, σ_{ik} is the standard deviation of processing time for operation k of product i and ε is the coefficient of standard deviation.

4.3 Capacity requirements in the presence of stochastic demand and processing times

In the deterministic case, the capacity requirement of a product is calculated via multiplying its demand with processing time. However, in the stochastic case, since both demand and processing time are probabilistic, the product of these two random variables becomes probabilistic and requires statistical analysis to find the probability density functions (pdf) of the capacity requirements. To find the fitted distribution (pdf) of the capacity requirement of product i, statistical analysis is performed with Arena Input Analyzer software.

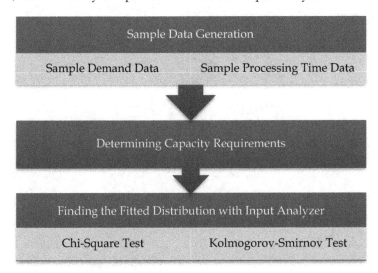

Fig. 1. The Framework of Input Analysis

The framework of the analysis is illustrated in Figure 1. Sample probabilistic demand and processing time data are generated. Capacity requirements are determined from the samples. The fitted distributions for capacity requirements are obtained from Input Analyzer software with respect to test results.

4.4 Stochastic capacitated non-linear cell formation

A stochastic non-linear mathematical model is developed by modifying Süer et al.'s (2010) deterministic model. The proposed model considers the variation of capacity requirements along with the mean capacity requirements. Product families and cell formations are determined with respect to available cell capacity and similarity coefficients. The indices, parameters and decision variables are listed as follows.

Indices:

i Product index
j Product index and family/cell index

Parameters:

S_{ij} Similarity coefficient between product i and j
μ_{CR_i} Mean capacity requirement for product i
σ_{CRi}^2 Variance of capacity requirement for product i
n Number of products
TU Upper limit for cell capacity
α Design factor

Decision Variables:

X_{ij} 1, if product i is assigned to family j ; 0, otherwise

Objective Function:

$$\max Z = \sum_{i=1}^{n}\sum_{j=1}^{n} S_{ij} * X_{ij} - \sum_{j=1}^{n} X_{jj} \tag{5}$$

Subject to:

$$p\left(Z_{nj} \leq \frac{\left(TU - \sum_{i=1}^{n}\mu_{CR_i} * X_{ij} \right)}{\sqrt{\sum_{i=1}^{n}\sigma_{CRi}^{2} * X_{ij}}} \right) \geq (1-\alpha) \quad j=1,2,\dots,n \tag{6}$$

$$\sum_{j=1}^{n} X_{ij} = 1 \qquad\qquad i = 1,\dots,n \tag{7}$$

$$X_{ij} \leq X_{jj} \qquad\qquad j = 1,\dots,n \, and \, i = 1,\dots,n \tag{8}$$

$$X_{ij} \in [0,1]; X_{ij} \, integer \tag{9}$$

The objective function is shown in equation 5. It maximizes the total similarity among products that are formed as families to be produced in dedicated cells, while minimizing the total number of cells. Equation 6 is the non-linear constraint which limits the cell utilization up to the cell capacity by considering mean and variance of capacity requirements based on a factor, α, which indicates the maximum acceptable probability that capacity requirements will exceed the capacity available. Equation 7 forces all products to be assigned to a cell. Equation 8 guarantees the assignment of each product to only one of the cells that are open. Equation 9 determines whether product i is assigned to cell j or not.

4.5 Simulation

In this study, the proposed solution methodology is validated with a simulation model. Even though, the CMS design literature is abundant with several mathematical models, model validation is considered in only a handful of works. Indeed, model validation is one of the significant requirements of any model-based solution methodology. Especially in a system where demand, processing times and capacity requirements are probabilistic, it is a must to validate the proposed approach. The type of model proposed in the study is white-box (causal descriptive) according to the Barlas's classification (Barlas, 1996). Therefore, it is expected that the model reproduces the behavior of the system studied. The behavior of the system is analyzed with respect to four measures, namely: cell utilization, WIP, waiting time and the number waiting. The hierarchical framework followed through the validation is shown in Figure 2.

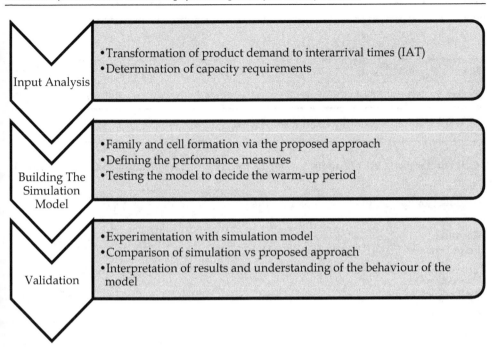

Fig. 2. The hierarchical framework

4.6 Example problem

An example problem is derived to explain the solution methodology (Please refer to the first 10 parts of part-machine matrix in Süer et al., 2010). There are 10 products in the system. The hybrid similarity matrix is shown in Table 8 and probabilistic capacity requirements are shown in Table 9. The capacity of a cell is considered as 800 hours for the example problem.

Part \ Part	1	2	3	4	5	6	7	8	9	10
1	1	0.808	0.765	0.555	0.849	0.561	0.75	0.68	0.712	0.636
2	0.808	1	0.685	0.718	0.747	0.59	0.677	0.775	0.631	0.597
3	0.765	0.685	1	0.545	0.794	0.569	0.703	0.597	0.751	0.627
4	0.555	0.718	0.545	1	0.539	0.562	0.545	0.681	0.477	0.636
5	0.849	0.747	0.794	0.539	1	0.605	0.692	0.65	0.656	0.611
6	0.561	0.59	0.569	0.562	0.605	1	0.499	0.69	0.507	0.656
7	0.75	0.677	0.703	0.545	0.692	0.499	1	0.583	0.855	0.542
8	0.68	0.775	0.597	0.681	0.65	0.69	0.583	1	0.534	0.715
9	0.712	0.631	0.751	0.477	0.656	0.507	0.855	0.534	1	0.499
10	0.636	0.597	0.627	0.636	0.611	0.656	0.542	0.715	0.499	1

Table 8. Hybrid similarity matrix

Part	1	2	3	4	5	6	7	8	9	10
Mean	154	153	494	93.9	106	500	138	45.6	439	135
Variance	231.04	262.44	5745.64	88.92	136.89	5343.61	187.69	25.1	6052.84	222.01

Table 9. Probabilistic capacity requirements (hrs)

Cell Formation	Cell 1 / Family 1	Cell 2 / Family 2	Cell 3 / Family 3	Cell 4 / Family 4
Product Family	Products (1,2,4,8)	Products (3, 5)	Products (6, 10)	Products (7, 9)
Resource Requirements	M/C (1-7,9,10,18)	M/C (1-4,8,10-12,18)	M/C (4,6,9,10,12,14,16, 18)	M/C (1,2,3,11,13,17, 18)
Expected Utilization	447	600	635	577
Util. w.r.t. 10 % risk	478 (60 %)	698 (87 %)	730 (91 %)	678 (85 %)

Table 10. Solution of example problem

The example problem is solved with the proposed non-linear mathematical model. The results are shown in Table 10. Products are formed as four families thus four cells are required where each is dedicated to one product family. Cell and family formations are the same since independent dedicated cells are assumed and intercell movements are not allowed.

5. Alternative solution methodology: Genetic algorithms

Genetic Algorithms (GA) is one of the most powerful metaheuristics used to solve NP-hard problems. It is usually used for solving large size problems where mathematical models run into computational and memory problems. The framework of GA is shown in Figure 3. This framework also represents the one cycle evolutionary process of GA.

GA consists of following steps (Süer, Mese, & Eğilmez, 2011):

1. Initial population of n chromosomes is formed randomly.
2. Mates are determined using the mating strategy to perform crossover.
3. The crossover and mutation operations are performed to generate offspring.
4. For selection, parents are added to the selection pool along with offspring.
5. The next generation is selected from this pool based on their fitness function values.
6. These steps are repeated until the number of the generations specified by the user is reached.
7. Finally, the best chromosome obtained during the entire evolutionary process is taken as the final solution.

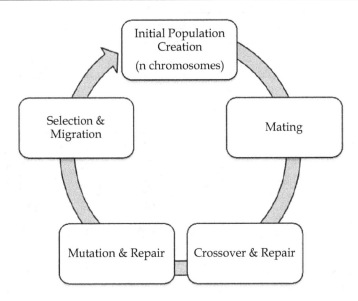

Fig. 3. Illustration of one generation

5.1 Initial population generation

Initial population is randomly generated based on the pre-defined number of chromosomes to form the population. Chromosome represents a candidate solution to the problem. In the proposed GA, the chromosome representation is designed to include product and cell numbers. An example chromosome is shown in Figures 4a and 4b. Each gene carries two types of information: part number and cell number. When generating chromosome, the product numbers are randomly assigned to genes as shown in Figure 4a. Once the assignment is finished the allocation of cells is done with respect to available capacity as shown in Figure 4b. The final chromosome representation is given in Figure 4b.

Cells are matched with products from left to right. First of all, cell 1 is opened and products are assigned to cell 1 as long as the cell capacity is available. If a product is going to result in overutilization, a new cell is opened. Product and cell allocation are illustrated in Table 11.

For example, after the assignment of products 1, 2 and 4 to cell 1, the total expected utilization of cell 1 is 400.9 hours and the standard deviation is 24.13. Under 10 % of risk ($z=1.28$), the upper bound of utilization is 400.9 + 1.28*24.13 = 431.79. If part 3 were to be assigned to cell 1, the total expected utilization would be 894.9 and variance would be 6328.04. The utilization under 10 % is equal to 894.9 + 1.28 * 79.55 = 996.72 hours > 800 hours. Since the cell is over utilized, a new cell is opened and product 3 is assigned to cell 2. The fitness function of a chromosome is the total similarity of cells. The similarity of cell is calculated via summing the similarities of products within the cell. The resulted cell formation and utilizations are shown in Table 11.

| 1 | 2 | 4 | 3 | 5 | 10 | 7 | 9 | 6 | 8 |

Fig. 4a. Example chromosome after product assignment

| 1,1 | 2,1 | 4,1 | 3,2 | 5,2 | 10,2 | 7,3 | 9,3 | 6,4 | 8,4 |

Fig. 4b. Example chromosome after cell allocation

Cell 1			Cell 2		
Part	μ	σ^2	**Part**	μ	σ^2
1	154	231.04	3	494	5745.64
2	153	262.44	5	106	136.89
4	93.9	88.92	10	135	222.01
Total	400.9	582.40	Total	735	6104.54
Cell 3			**Cell 4**		
Part	μ	σ^2	**Part**	μ	σ^2
7	138	187.69	6	500	5343.61
9	439	6052.84	8	45.6	25.10
Total	577	6240.53	Total	545.6	5368.71

Table 11. Cell utilization (hrs)

5.2 Mating strategy

Random Mating Strategy is used in mating. Firstly, the reproduction probabilities of the chromosomes are calculated according to their fitness function. Each chromosome in the population is mated with a randomly selected partner and they produce one offspring. The partner is selected using reproduction probability based on Roulette Wheel approach.

5.3 Crossover operation and repair

A modified order crossover method is employed for the problem studied. In the order crossover as represented in Figure 5, a random number between 1 and the maximum number of cells (4, in the example given in Figure 5) is drawn. Assume that 2 is drawn. This is to decide how many cells from parent 1 are going to be kept in offspring. Then, the two cells are randomly selected and copied to offspring. The remaining genes are filled from parent 2 based on the order of remaining products in parent 2's chromosome. Finally, cell numbers of products that are obtained from parent 2 are assigned based on the availability of cell capacity. As long as the cell capacity is available, products from left to the right are assigned to the same cell. If a capacity violation occurs, a new cell is opened.

5.4 Mutation operation and repair

Mutation is only applied to part numbers and after the mutation operation, same repair operation as in crossover strategy is employed to re-identify cell formation based on available cell capacity. Each gene is mutated subject to a mutation probability. If a mutation occurs in a gene, a random part number is replaced with an existing part number in a particular cell.

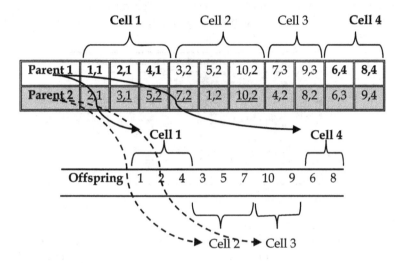

Fig. 5. Cell-Based Order Crossover (CBOC)

5.5 Selection and migration

Selection pool consists of all offspring and parents. The best chromosome is selected as the best solution for the particular generation. After each generation, a predetermined portion of existing generation is advanced to the next generation. Assume that x is the desired percentage of selection pool advancing to the next generation. X is basically an experimentation parameter. When generating next generation, the remaining positions are filled with randomly generated chromosomes (thus the term migration).

6. Experimentation and results

Experimentation consists of two sections. In the first section, the proposed stochastic non-linear mathematical model is compared with Süer et al.'s (2010) deterministic model. The results of both approaches are compared with respect to the obtained cell formation and simulation results. In the second section, proposed alternative solution methodology (GA) is compared with the non-linear mathematical model. Five different problems are considered, namely: 10, 20, 30, 45 and 60 parts. In the first section, the 30-part problem is used for the comparison and simulation runs. For the comparison of non-linear mathematical model with GA, all five problems are used with respect to the solution quality and execution time.

6.1 Data generation

The part-machine matrix (30x18) is obtained from a jewelry manufacturing company. Processing times are generated from uniform distribution with (15, 25) minutes. Each cell is independent, i.e. part undergoes all the operations in only one cell and comes out as a final product. Therefore, intercell movement of parts is not allowed. Independent cells are used in certain systems where intercell movement of parts is either not possible (pharmaceutical manufacturing) or may cause serious problems due to product mix up. One piece flow principle is assumed for the entire cellular manufacturing system. Set-up times for the parts are assumed as zero.

Demand Category	Annual Uniform Demand Distribution	
	Lower Bound	Upper Bound
1	250	750
2	751	1250
3	1251	1750
4	1751	2250
5	2251	2750

Table 12. Annual Uniform Demand Data Generation

Cell capacity is assumed as 2000 hours (50 weeks x 40 hrs per week) per year. Remaining two weeks is allocated to compensate for unexpected system breakdowns and plant shutdowns. Demand for each product is assumed to follow normal distribution. The mean demand is generated from uniform distribution from five categories. Each product is assigned randomly to a category (see Table 12). The standard deviation of demand is generated via multiplying the mean demand with a factor. The factor is obtained randomly via uniform distribution (0.1, 0.5).

6.2 Comparison of Süer et al.'s (2010) model with the proposed stochastic approach

In this section, the results of Süer et al.'s (2010) deterministic approach (see Figure 6) and the proposed stochastic approach (see Figure 7) are provided. The deterministic model grouped the products into 10 families/cells based on the hybrid similarity matrix. Simulation experiment resulted in 100% utilization in cell number 5 and second highest utilization is observed in cell number 2. Since deterministic model only considers mean capacity requirements of products, some of the cells are utilized over 95% when deterministic mathematical model is used. However, these high utilization rates resulted in the same or lower utilization with simulation experimentation. The overall trend of utilization obtained from simulation is observed as similar to the result of mathematical model.

According to the results of the proposed stochastic non-linear approach (Figure 7), products are formed as 13 families/cells. The proposed approach increased the number of cells by 3 and the number of machines by 18%. A similar trend between the result of mathematical model and simulation is also observed with the stochastic approach. In contrast to very high utilization observed on 2 cells in the deterministic approach, the highest utilization is obtained as 82% with the stochastic approach (from simulation). Simulation model resulted in 1% to 7% less utilization of cells than mathematical model's results. There is no overutilization observed in any cell with simulation since the proposed approach considers variance of capacity requirements in addition to similarities during the cell formation.

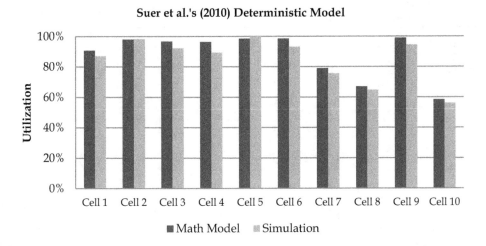

Fig. 6. Utilization Results of Deterministic Model by Süer et al. (2010)

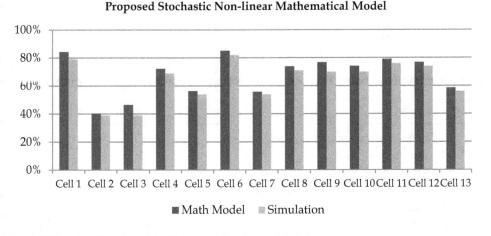

Fig. 7. Utilization Results of the Proposed Stochastic Model

Utilization-based comparison is important to observe the model validation with simulation experiments. However, the behavior of model is also important in model validation. Therefore, system performance is also included in comparison. The performance measures considered are 1) cell utilization, 2) the number of machines, 3) work-in-process (WIP) inventory, 4) average waiting time and 5) average number waiting. The results obtained from both approaches are shown in Table 13.

According to the system performance comparison (Table 13), WIP, average waiting time and average number of waiting decreased significantly since there is no overutilization occurred with stochastic approach. However, the number of machines increased from 94 to 111.

Cell	# M/C	WIP	Av. Waiting Time	Av. Number Waiting	Cell	# M/C	WIP	Av. Waiting Time	Av. Number Waiting
Deterministic Approach (Süer et al., 2010)					Proposed Stochastic Approach				
1	10	6.06	18.8	1.22	1	17	3.53	11.73	0.44
2	13	7.84	12.17	1.05	2	10	2.27	2.09	0.06
3	10	6.06	19.11	1.09	3	9	1.64	2.72	0.11
4	12	6.62	17.57	0.97	4	8	1.72	2.50	0.14
5	13	298.46	6256.56	294.30	5	7	5.00	5.42	0.18
6	12	6.83	18.45	1.03	6	11	6.02	9.90	0.43
7	4	2.52	1.04	0.03	7	11	5.53	4.73	0.22
8	8	3.9	0.39	0.01	8	6	4.37	3.98	0.30
9	8	5.72	21.37	1	9	8	2.95	5.24	0.26
10	4	1.94	0.07	0	10	9	7.60	7.90	0.42
Total	94	345.97	6365.53	300.71	11	4	1.90	1.02	0.03
					12	7	3.86	7.52	0.30
					13	4	1.95	0.07	0.00
					Total	111	48.34	64.82	2.90

Table 13. System Performance Based Comparison

6.3 Comparison of the proposed stochastic non-linear mathematical model with GA

The problem studied is NP-Hard. Therefore, it is important to provide alternative solution approaches that are providing near optimal solution in faster times when dealing with larger problems. In this section, proposed mathematical model and genetic algorithms are compared. The comparison is made in two ways: 1) system performance and 2) solution quality and execution time. Five datasets are used consisting of 10, 20, 30, 45 and 60 parts. Mathematical model could only provide global optimal solutions for the first 4 datasets. To be able to analyze system performance, both solutions for dataset with 30 parts obtained from mathematical model and GA are simulated. The simulation results are shown in Table 14. According to the results, both approaches formed 30 products as 13 families and opened 13 dedicated cells. The utilizations of cells vary from 39% to 82%. In terms of total number of machines, GA resulted in 10 less machines. Even though, there is parallelism between WIP and average number waiting results, in terms of average waiting times, GAs solution provided more than 50% lower average waiting time. However, both mathematical model and GA results showed better performance than Süer et al.'s (2010) deterministic model.

Secondly, both approaches are compared based on the solution quality and execution time. The optimal solution of mathematical model, and best and average solution of GA, execution times and the average % distance (gap) GA's solution from mathematical model and parameter set used in GA are shown in Table 15. Mathematical model provided global optimal solution for the first 4 datasets. On the other hand, GA provided the optimal solution with 100% frequency with 10 parts dataset. With the datasets: 20, 30 and 45 parts; GA provided near optimal solutions with great improvements in execution times. Since the solutions of 30 parts dataset provided by GA and mathematical models are compared in detail in Table 14, it can be concluded that near optimal solutions with respect to similarity coefficient can still provide good or even better system performance. However, system performance should always be assessed with validation methods e.g. simulation.

Cell	Stochastic Non-linear Mathematical Model				Genetic Algorithms			
	# M/C	WIP	Av. Waiting Time	Av. Number Waiting	# M/C	WIP	Av. Waiting Time	Av. Number Waiting
1	17	3.53	11.73	0.44	8	0.94	2.93	0.11
2	10	2.27	2.09	0.06	6	2.39	2.03	0.15
3	9	1.64	2.72	0.11	5	2.46	1.85	0.11
4	8	1.72	2.50	0.14	5	0.13	0.04	0.00
5	7	5.00	5.42	0.18	4	1.41	0.51	0.02
6	11	6.02	9.90	0.43	8	0.28	0.03	0.00
7	11	5.53	4.73	0.22	10	1.92	3.00	0.12
8	6	4.37	3.98	0.30	14	6.15	5.43	0.34
9	8	2.95	5.24	0.26	8	4.57	1.48	0.11
10	9	7.60	7.90	0.42	6	5.48	0.46	0.02
11	4	1.90	1.02	0.03	7	1.87	0.20	0.01
12	7	3.86	7.52	0.30	12	14.54	3.58	0.15
13	4	1.95	0.07	0.00	8	5.73	4.25	0.20
Total	111	48.34	64.82	2.90	101	47.86	25.79	1.34

Table 14. System Performance Comparison of Mathematical Model and GA

Problem Size	Stochastic Non-linear Mathematical Model		Genetic Algorithms				
	Optimal Solution	Sol. Time (Sec)	Best Solution	Avr. Solution	Avr. % Gap	Avr. Exe. Time (Sec)	Parameter Set (IP,NOG,CP,MP,MR)
10	4.606	145	4.606	4.606	0%	9.70	(100, 200, 0.7-0.9, 0.01, 10-20%)
20	10.87	1522	10.57	10.11	7%	16.90	(500, 100, 0.7-0.9, 0.01, 10-20%)
30	12.29	40836	11.80	11.53	6%	32.93	(1000, 1000, 0.7-0.9, 0.01, 10-20%)
45	20.064	198662	17.024	16.315	19%	78.13	(1000, 2000, 0.7-0.9, 0.01, 10-20%)
60	N/A	N/A	26.855	25.164	-	114.31	(1000, 2000, 0.7-0.9, 0.01, 10-20%)

IP: Initial population, NOG: Number of generations, CP: Crossover probability, MP: Mutation probability, MR: Migration rate

Table 15. Comparison Based on Solution Quality and Execution Time

7. Conclusion

In this study, the impact of probabilistic demand and processing times on cell formation is addressed. The uncertainty in demand and processing times is one of the most common problems of manufacturing world. Manufacturing system design is also directly influenced by such factors which may result in million dollars of wrong investment on machines and equipments. Majority of literature on cell formation and manufacturing system design either neglects the uncertainty in demand and processing times or assumes the impact as limited. Süer et al. (2010) proposed two approaches; deterministic and stochastic. Deterministic model used in their study formed the cells based on the expected (mean) capacity requirements. On the other hand, stochastic approach proposed a layered manufacturing system design to deal with uncertainty in demand and processing times. They allowed a family to have more than one cell to deal with uncertainty. In this study, each family is restricted to one cell and the impact of variance is included in the proposed approach. A hierarchical methodology is used to solve the problem. First of all a new similarity coefficient is introduced which incorporates the route and demand similarity in one similarity coefficient, "hybrid similarity coefficient". Hybrid similarity matrix is built based on the data obtained from a jewelry manufacturing company where the operations are labor intensive. To deal with the cell and family formation under the impact of uncertain demand and processing times, a stochastic non-linear mathematical model is developed.

The proposed model and Suer et al.'s (2010) deterministic model are experimented with 30x18 (part x machine) dataset. During the modeling and experimentation phases, independent cell and family formation is considered. Therefore, intercell movement of parts is not allowed. Each formed product family is assumed to be produced by a dedicated cell. In addition to solving the cell and family formation problem with mathematical optimization, model validation is also considered. To validate the designed cellular manufacturing system, simulation models are developed for both proposed stochastic and Süer et al's (2010) deterministic approaches. According to the simulation results, similar cell utilization patterns are obtained between the mathematical model and simulation results for both proposed and Suer et al.'s (2010) deterministic approaches. In the further analysis, both approaches are compared based on the system performance. Five performance indicators are considered in comparison, namely: cell utilization, the number of machines, WIP, average waiting time, average number waiting. According to the results, the proposed approach resulted in more number of cells and more machines, but it performed better with respect to all other performance measures.

Even though, the proposed non-linear stochastic model with respect to hybrid similarity coefficient guarantees the optimal solution, solution time increases exponentially as the number of parts increases due to the NP-hardness of problem studied. Metaheuristics are usually employed to deal with NP-hard problems. Therefore, a Genetic Algorithm (GA) model is also developed to deal with the medium and large scaled problems. The probabilistic parameters are also reflected in GA. Experimentation is performed with five datasets, namely 10, 20, 30, 45 and 60 parts. All other datasets are generated as a portion or random replication of 30 parts datasets. First of all, the solution obtained from GA for 30 parts is simulated. Both stochastic non-linear mathematical model and GA's simulation results are compared. According to the simulation results, GA performed better in terms of all performance measures. In addition, both GA and mathematical model resulted in better performance than the deterministic model. Secondly, remaining datasets are also experimented with both proposed mathematical model and GA. According to the results, GA found the optimal solution with 100% frequency for the first dataset (10 parts). For the larger datasets 20, 30 and 45 parts, GA provided 6% to 19% distant solution than the optimal solution. The largest (60 parts) dataset can only be solved by GA. Besides, GA significantly outperformed the mathematical model in all datasets in terms of execution time. Another important conclusion is that the optimal solution with respect to similarity does not guarantee that system performance will be better. The proposed CMS is required to be validated with simulation.

Even though GA provided good and faster solutions and to the best knowledge of authors there has not been any GA approach including stochastic components proposed for the problem studied in literature yet, the proposed GA is planned to include alternative genetic operators to be able to increase the solution quality. In addition, the problem studied can be extended to include other features of CMS design such as allowance of intercell movement of parts, system implementation costs, setup times etc. The impact of variance on system design is another potential important side of the problem which may affect the solution significantly. Moreover, identification of the bottleneck machine in systems where bottleneck machine shifts is also another important potential future work. Finally, in addition to genetic algorithms, other meta-heuristics such as Simulated Annealing, Tabu Search can be considered to cope with the larger problems.

8. References

Akturk M.S., & Turkcan A. (2000). Cellular manufacturing system design using a holonistic approach. *International Journal of Production Research, 38*(10), 21. Taylor and Francis Ltd, ISSN: 0020754.

Albadawi, Z., Bashir, H. A., & Chen, Mingyuan. (2005). A mathematical approach for the formation of manufacturing cells. *Computers and Industrial Engineering, 48*(1), 3 - 21, ISSN: 03608352.

Asokan, P., Prabhakaran, G., & Satheesh Kumar, G. (2001). Machine-Cell Grouping in Cellular Manufacturing Systems Using Non-traditional Optimisation Techniques - A Comparative Study. *The International Journal of Advanced Manufacturing Technology, 18*(2), 140-147. Springer London, ISSN: 0268-3768.

Barlas, Y. (1996). Formal aspects of model validity and validation in system dynamics. *System Dynamics Review, 12*(3), 183-210, ISSN: 08837066

Cao, D., & Chen, M. (2004). Using penalty function and Tabu search to solve cell formation problems with fixed cell cost. *Computers & Operations Research, 31*(1), 21-37, ISSN: 03050548.

Chen, Mingyuan. (1998). *A mathematical programming model for system reconfiguration in a dynamic cellular manufacturing environment. Annals of Operations Research* (Vol. 77, pp. 109-128-128). Springer Netherlands, DOI: 10.1023/A:1018917109580.

Harhalakis, G., Nagi, R., & Proth, J. M. (1998). An Efficient Heuristic in Manufacturing Cell Formation for Group Technology Applications. *International Journal of Production Research, 28*(1), 185-198, DOI: 10.1080/00207549008942692.

Heragu, S., & Chen, J. (1998). Optimal solution of cellular manufacturing system design: Benders' decomposition approach. *European Journal of Operational Research, 107*(1), 175-192, ISSN: 03772217.

Jayaswal, S., & Adil, G. K. (2004). Efficient algorithm for cell formation with sequence data, machine replications and alternative process routings. *International journal of production research, 42*(12), 2419-2433. Taylor & Francis, ISSN: 0020-7543.

Kamrani, A. K. A. K., Parsaei, H. R., & Leep, H. R. (1995). *Planning, Design, and Analysis of Cellular Manufacturing Systems. Manufacturing Research and Technology* (Vol. 24, pp. 351-381). Elsevier, ISBN10: 0-444-81815-4

Kusiak, A. (1987a). The generalized group technology concept. *International Journal of Production Research, 25*(4), 561-569, ISSN: 0020-7543.

Kusiak, A. (1987b). The generalized group technology concept. *International Journal of Production Research, 25*(1), 561-569m, ISSN: 01663615.

McAuley, J. (1972). Machine grouping for efficient production. *Production Engineer, 51*(2), 53-57, ISSN: 0032-9851 .

Moon, C., Gen, M., & Suer, G. A. (1999). A genetic algorithm-based approach for design of independent manufacturing cells. *International Journal of Production Economics, 60-61*(2), 421-426, ISSN: 09255273.

Purcheck, G. F. K. (1974). A mathematical classification as a basis for the design of group-technology production cells. *Production Engineer, 54*(1), 35-48, ISSN: 0032-9851.

Rajamani, D., Singh, N., & Aneja, Y. P. (1990). Integrated design of cellular manufacturing systems in the presence of alternative process plans. *International journal of production research, 28*(8), 1541-1554. Taylor & Francis, ISSN: 0020-7543.

Saad, S. (2003). The reconfiguration issues in manufacturing systems. *Journal of Materials Processing Technology, 138*(1-3), 277-283, ISSN: 09240136.

Seifoddini, H. (1990). A probabilistic model for machine cell formation. *Journal of Manufacturing Systems, 9*(1), 69-75, ISSN: 02786125.

Shafer, S., & Rogers, D. (1991). A goal programming approach to the cell formation problem. *Journal of Operations Management, 10*(1), 28-43, ISSN: 02726963.

Shtubt, A. (1989). Modelling group technology cell formation as a generalized assignment problem. *International Journal of Production Research, 27*(5), 775 - 782. Taylor & Francis, ISSN: 00207543.

Silver, E. A., & Peterson, R. (1985). *Decision Systems for Inventory Management and Production Planning (Wiley Series in Production/Operations Management)* (p. 736). Wiley, ISBN: 0471867829.

Sofianopoulou, S. (1999). Manufacturing cells design with alternative process plans and/or replicate machines. *International Journal of Production Research, 37*(3), 707-720. Taylor & Francis, ISSN: 0020-7543.

Solimanpur, M., Vrat, P., & Shankar, R. (2004). A multi-objective genetic algorithm approach to the design of cellular manufacturing systems. *International Journal of Production Research, 42*, 1419-1441, ISBN: 978-1-4244-3334-6.

Suer, G. A., Huang, J., & Sripathi, M. (2010). Design of dedicated, shared and remainder cells in a probabilistic demand environment. *International journal of production research, 48*(19-20), 5613-5646. Taylor & Francis, ISSN: 0020-7543

Suer, G. A., Pena, Y., & Vazques, R. (2003). Cellular design to minimise investment costs by using evolutionary programming. *International Journal of Manufacturing Technology and Management, 5*(1), 118-132, DOI: 10.1504/IJMTM.2003.002536

Süer, G. A., Mese, E. M., & Eğilmez, G. (2011). Cell Loading, Family and Job Scheduling to minimize TT. *International Conference on Intelligent Manufacturing Logistics Systems, February 27-March 1, 2011, Chung-Li, Taiwan.*

Wang, J. (1998). A linear assignment algorithm for formation of machine cells and part families in cellular manufacturing. *Computers & Industrial Engineering, 35*(1-2), 81-84, ISSN: 03608352

Wei, J. C., & Gaither, N. (1990). An Optimal Model for Cell Formation Decisions. *Decision Sciences, 21*(1), 416-433.

Wemmerlov U., & Johnson D. J. (1997). Cellular manufacturing at 46 user plants: implementation experiences and performance improvements. *International Journal of Production Research, 35*(1), 29-49. Taylor and Francis Ltd, DOI: 10.1080 /002075497195966

Wicks, E. M., & Reasor, R. J. (1999). *Designing cellular manufacturing systems with dynamic part populations*. *IIE Transactions* (Vol. 31, pp. 11-20-20). Springer Netherlands, DOI: 10.1023/A:1007516532572.

Stochastic Multi-Stage Manufacturing Supply Chain Design Considering Layered Mini-Cellular System Concept

Jing Huang and Gürsel A. Süer*
Ohio University
USA

1. Introduction

Supply chain design attempts to deploy resources to synchronize product flow through multiple tiers of the network and eventually fulfill customers' requirements (Lee, 2000). Traditionally, due to limited sales information and trade barriers, consumers chose local products. But nowadays consumers access the ever opening global market much easier with the help of the globalization. Though overall demand keeps increasing, the market competition becomes fierce as more and more competitors emerge. The highly dynamic market results in difficulties in predicting the demand for companies' products. Furthermore, the demand uncertainties exacerbate the challenge for synchronizing production. Companies start to build more safety stock and hold excess capacity, resulting in decrease of system efficiency. Thus, resource allocation becomes a critical part in the supply chain design. Three levels of resource allocation in supply chain design are summarized in Figure 1. This is only an attempt to present multiple perspectives of resource allocation problems without guaranteeing the coverage of all industrial circumstances. The exceptions and variations in this framework could always be found in the real world applications.

As the market boosts globally, where to manufacture, store and sell various products become the first decision in resource allocation. "Where" could refer to the market such as North American and East Asian (market and production allocation). In this case, managerial decisions related to marketing strategy and business practice are involved. Some auto manufacturers such as Toyota and Honda open local manufacturing facilities in every market they enter. They take the advantage of local resources to increase responsiveness to local market. Furthermore, the risks caused by demand fluctuation are limited to individual markets and does not adversely affect other operations located throughout the world.

"Where" could also mean a specific geographical location for a specific facility in the supply chain network (facility location). By keeping products manufactured or stocked in one central place, company could benefit from economics of scale and increase its efficiency; however, they might reduce responsiveness. Locating resources dispersedly but close to the consumers could improve customer's satisfaction level, but this increases complexity in

---

* Corresponding Author

coordination of product flow. Though Honda and Toyota open facilities for each local market, their local manufacturing facilities are clustered within a certain area to facilitate just-in-time system thereby reducing leadtime tremendously. On the contrary, Seven-Eleven builds its supply facilities close to its convenience stores in Japan. Each store could efficiently manage its inventory by using the Total Information System, where each order is tracked and recorded by the scanner terminal. Distribution centers receive food from manufacturing plants and directly transfer it to the trucks instead of carrying any inventory for fast food. Thus, Seven-Eleven is able to provide fresh product such as lunch box, sandwiches, bakery and bread and improve its responsiveness (Chopra & Meindl, 2007).

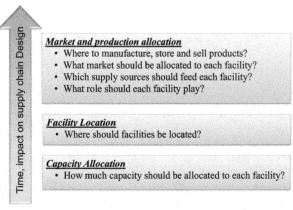

Fig. 1. Three main issues in supply chain design

In the supply chain operational phase, "where" refers to a specific capacity in the shop floor, warehouse, or transportation (capacity allocation). Quantitative models are applied to study machine capacity planning and transportation planning problems where machine or vehicle utilization could be optimized to meet the demand on time under various demand patterns.

Market and production allocation was discussed by Dicken and summarized as: "globally concentrated production", "host market production", "product-specialization for a global or regional market", and "transnational vertical integration" (Dicken, 1992). Globally concentrated production holds production in one base and ships products worldwide. No doubt, production cost could be reduced if production is located in a low-labor cost country; however, the risks of delayed response to the market change arise. Host market production eliminates this risk by dispersing production to each individual market without allowing sales across market boundaries. A better understanding of local customers could be developed and sensitivity to market change could be maintained in each individual market. In the third type of production location strategy, each of the markets manufactures only one product group that is sold to other markets as well. This strategy creates a large-scale and highly specialized manufacturing environment where production cost decreases but transportation cost increases. Transnational vertical integration strategy assigns components or semi-finished products to each of the markets based on manufacturing process, and eventually assemble finished products in one market. It takes advantage of geographical variation of production cost, especially the labor cost. For example, producing low-tech components in developing countries but core components in developed countries could minimize the cost while maintaining product quality. However, additional transportation cost is added if finished products are sold back to the market where components are manufactured.

Managerial decisions related to marketing strategy and business practice are involved in making appropriate market and production allocation decisions. Thus, this chapter assumes that "transnational vertical integration" strategy is adopted, where components are produced in various geographical locations and sold in North American market.

Facility location and capacity allocation decisions are then addressed and solved by a quantitative model in this chapter. Facility location and capacity allocation determines the location, allocation, and production/delivery volume of the flow of goods in a supply chain. Efficient use of all resources to handle supply chain uncertainties is important to facilitate supply chain coordination thereby improving companies' competitiveness (Lee, 2005).

Demand uncertainty is one of the main obstacles in making appropriate decisions. To satisfy demand, supply chain designer tends to reserve extra capacity; however this results in low utilization and therefore higher production cost. Reserving too little capacity results in demand shortage and low responsiveness.

In addition, supply chain design and manufacturing system design are traditionally two sequential steps (Rao & Monhanty, 2003; Cosner, 2008; Schaller, 2008). Roughly estimated capacity requirements are used to locate facility and allocate the production. Various manufacturing systems are then formed within each selected facility. Inaccuracy of estimated capacity requirements results in unsuitable supply chain design and further decreases the manufacturing performance in each facility.

We are proposing a four-phase approach to design and implement the layered mini-cellular system for a multi-stage manufacturing supply chain. In this study, the manufacturing system design and supply chain design are integrated into one scenario. A layered mini-cellular manufacturing system is adopted in the production facility, which is discussed in detail later. Mini-cells are first formed based on probabilistic demand. The mini-cell formation results then serve as inputs to a capacitated plant location model to determine which potential manufacturing plant is selected and how much capacity is allocated to this plant for each manufacturing stage. To continue studying on supply chain operational decisions, a production planning model is proposed to help decide detailed production quantity in each mini-cell for each manufacturing stage in each period.

The remainder of the chapter is organized as follows. The proposed layered mini-cellular system is introduced in section 2. In section 3, solution methodologies are discussed in detail. Experimentation results of proposed system are reported in section 4. In section 5, the performance of layered design is investigated and compared with a classical system. The conclusion is drawn in section 6.

2. Proposed layered mini-cellular system

The manufacturing systems can be categorized into fixed, product, process, and cellular layout in terms of its production layout. Fixed layout is particularly designed for heavy or fragile products such as airplanes, submarines and trains. The product stays in a fixed position, and machines are moved around the product to finish tasks. Product layout is usually adopted by the system with low product diversity but high volume. Each product line is designed for a specific product and performs very efficient production with short throughput time and low work-in-process inventory. On the other hand, process layout is appropriate for a system with high product variety but low volume. Similar processes/machines are grouped and shared by different products, therefore increasing utilization. However, the multidirectional production flow brings challenge in shop floor control. In a cellular layout, products are grouped into families based on the process similarities first and then produced in their own cells. Cellular layout integrates the essentials of both the product layout (product dedication) and process layout (process similarity) into one scenario. It is able to deal with high product variation, in the meanwhile, still maintain a relatively synchronized flow within each cell. The further advantages of cellular system include shorter set-up times, shorter leadtimes, less work-in-process inventory, and fewer defects.

In a classical cellular manufacturing system, each cell is dedicated to only one product family. The cell requirements may vary significantly under a highly fluctuating demand situation, which results in poor utilization of resources. Süer (Süer et al., 2010) brought more flexibility into the cellular system by introducing shared and remainder cells. Assume that the capacity requirements are computed based on the normally distributed demand and processing times (more detailed discussion is in section 3.1) as represented in Table 1. Both expected utilization of Xth cell and accumulated demand coverage by X cells are reported in Table 1. For instance, 0.009/0.999 implies that the 4th cell of family 1 is utilized 0.9% of the time, and four cells together could cover demand of family 1 99.9% of the time. To be able to cover the production demand 99.9% of the time, 4+3=7 cells are required for product families 1 and 2. It is clear that the 4th cell of family 1 and the 3rd cell of family 2 are rarely utilized. The demand of family 1 is still covered 99% of the time even without the 4th cell, thus, we may eliminate this cell. However, the 3rd cell of family 2 could not be avoided; otherwise the demand of family 2 will be only covered 86% of the time. In this case, we may group the 3rd cell of family 2 with the 3rd cell of family 1. The capacity requirement is reduced, in the mean time, the desired demand coverage (99%) for each product family is also guaranteed.

Expected Utilization / Demand Coverage	Family 1	Family 2
1st Cell	0.98/0.02	0.99/0.3
2nd Cell	0.9/0.39	0.84/0.86
3rd Cell	0.75/0.99	0.05/0.993
4th Cell	0.009/0.999	

Table 1. An example of capacity requirements

In the layered cellular system, a cell with poor utilization might be combined with another cell. In a shared cell, two product families can be processed. A remainder cell can handle more than two product families. Troubles caused by unstable demand are limited to 'shared' and 'remainder' mini-cells, and demand compensation effect among various product families could help to stabilize demand. The layered system is illustrated in Figure 2, where the production flow is assumed to be unidirectional in each cell. Considering that the chapter mainly studies the supply chain design, the batch production is assumed to simplify the capacity computations.

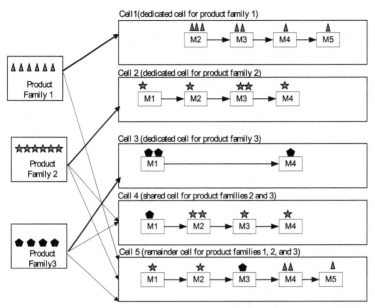

Fig. 2. Three main issues in supply chain design (adopted from Süer et al., 2010)

The layered cellular system proposed by Süer (Süer et al., 2010) assumes a single-stage manufacturing system. In the real-world applications, a multi-stage manufacturing system is usually involved in the manufacturing tier of a supply chain, where each manufacturing facility only performs partial production. In this chapter, 'cell' concept is evolved to a 'mini cell' concept. A cell performs full package production, while a mini-cell performs operations in a specific manufacturing stage.

3. Stochastic multi-stage manufacturing supply chain design

A four-phase approach is proposed to design and implement the layered mini-cellular system for solving resource management problem in a multi-stage manufacturing system. Table 2 summarizes phases and methodologies used in each phase.

Phase	Objective	Solution Method
1. Expected mini-cell utilization determination	Computing the number of mini-cells and their expected utilizations for each manufacturing stage	Probability Theory
2. Mini-cell formation	Grouping mini-cells for each manufacturing stage	Heuristic Procedure
3. Supply chain network design	Selecting production facilities and allocating mini-cells to the selected facilities	Mixed Integer Linear Programming
4. Simulation	Multi-period production planning	Mixed Integer Linear Programming

Table 2. Summary of stochastic multi-stage manufacturing supply chain design

3.1 Determining expected mini-cell utilization

In this phase, mini-cell requirement for each manufacturing stage is computed based on product demand and processing times. Mean and standard deviation of capacity requirements (in hours) for product family i is determined by the processing time (in minutes) of bottleneck operation in manufacturing stage j shown in Equations 1 and 2, where N_i is the number of parts in product family i, $\mu_{Deamand_{in}}$ is the demand mean of part n in product family i, $\sigma_{Deamand_{in}}$ is the demand standard deviation of part n in product family i, and $PTBottleneck_{ijn}$ is the processing time of bottleneck machine for part n in product family i at stage j.

$$\mu_{CRij} = \sum_{n=1}^{N_i} \left(\mu_{Demand_{in}} \times PTBottleneck_{ijn} / 60 \right) \tag{1}$$

$$\sigma_{CRij} = \sqrt{\sum_{n=1}^{N_i} \left[\left(\sigma_{Demand_{in}} \times PTBottleneck_{ijn} / 60 \right)^2 \right]} \tag{2}$$

Each mini-cell is assumed to work 40 hours per week. Thus, the probability of covering the demand by a mini-cell is computed as shown in Equation 3, where $X^{th}C_{ij}$ implies the Xth mini-cell required by product family i in stage j.

$$P(X^{th}C_{ij}) = cdf_{Normal}((40 \times X - \mu_{CRij}) / \sigma_{CRij}) \tag{3}$$

The expected utilization of the Xth mini-cell for product family i in stage j is given in Equation 4. $P(NCR_{ij}>X)$ indicates the probability that the number of mini-cells required by product i in stage j is greater than X as given in Equation 5, while $P(X\text{-}1<=NCR_{ij}<=X)$ means the probability that mini-cells required is between $X\text{-}1$ and X as given in Equation 6. PU_1 is the utilization of Xth mini-cell when mini-cell requirement is greater than X, therefore it is fixed as 1. PU_2 is computed as given in Equation 7, where μ_i is the mean demand of product family i, and σ_i is the standard deviation of the demand.

$$E(X^{th}C_{ij}) = P(NCR_{ij} > X) \times PU_1 + P(X - 1 \le NCR_{ij} \le X) \times PU_2 \tag{4}$$

$$P(NCR_{ij} > X) = 1 - P(X^{th}C_{ij}) \tag{5}$$

$$P(X - 1 \le NCR_{ij} \le X) = \begin{cases} P(X^{th}C_{ij}) & X = 1 \\ P(X^{th}C_{ij}) - P((X-1)^{th}C_{ij}) & X \ne 1 \end{cases} \tag{6}$$

$$PU_2 = \int_{40(X-1)}^{40X} \frac{y \times \frac{1}{\sigma_i \sqrt{2\pi}} e^{-(y-\mu_i)^2 \frac{1}{2\sigma_i^2}}}{40 \times P(X-1<NCR_{ij}<X)} dy - (X-1) \tag{7}$$

An example result of mini-cell capacity estimation for manufacturing stage j is shown in Table 3. The results imply that 4+4+3=11 mini-cells are required to cover the demand of these three product families. The expected utilization and demand coverage of each mini-cell are also given in the same table. Please note that, we will continue to use this small example to illustrate the procedures that will be discussed in the following sections.

Expected Utilization / Demand Coverage	Family 1	Family 2	Family 3
1st Mini-Cell	0.99/0.002	0.99/0.5	0.98/0.6
2nd Mini-Cell	0.87/0.39	0.79/0.86	0.55/0.91
3rd Mini-Cell	0.32/0.96	0.1/0.98	0.02/0.999
4th Mini-Cell	0.03/0.999	0.003/0.999	

Table 3. An example of mini-Cell utilization and demand coverage

3.2 Grouping mini-cells

In Table 3, obviously, several mini-cells are rarely utilized (e.g. 0.3% utilization of fourth mini-cell for family 2). A heuristic procedure is introduced in this section to reduce the number of required mini-cells by grouping mini-cell segments. The grouping process is implemented based on process similarities among product families. Another important criterion of the grouping process is demand coverage. For example, in Table 3, three mini-cells are able to cover the demand for product family 1 96% of the time, therefore, the fourth mini-cell might not be required.

Figure 3 illustrates the heuristic procedure, where XC_i implies X^{th} mini-cell for product family i, XCU_i is the utilization of this mini-cell, C_j is the newly formed mini-cell j, and LC_j is the leftover utilization for newly formed mini-cell j. Heuristic procedure attempts to form dedicated, shared and remainder mini-cells with the objective of reducing the total number of mini-cell requirements. In the meantime, it prefers to group product families with similar manufacturing operations in order to avoid increasing machine/workforce numbers and operational complexities within a mini-cell. This heuristic procedure is repeated for each manufacturing stage.

An example result of grouping 11 dedicated mini-cells (see Table 3) for manufacturing stage j is shown in Figure 4. After the grouping procedure is applied, four dedicated mini-cells stay, and two dedicated mini-cells are grouped into a shared mini-cell. The other three mini-cells originally dedicated to product families 1, 2, and 3 are grouped into a single remainder mini-cell. Since three mini-cells are able to cover the demand for product family 1 96% of the time, the fourth mini-cell required by product family 1 (noted in the red block) is abandoned during the grouping process. The same procedure is applied to the fourth mini-cell required by product family 2. It is observed that 11 mini-cells cover the demand all the time, and the grouping process reduces the number of mini-cells to six still covering demand 96% of the time.

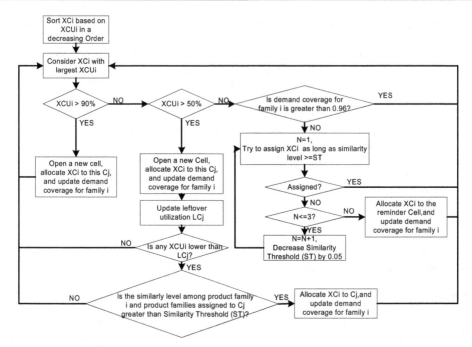

Fig. 3. Flowchart of heuristic procedure

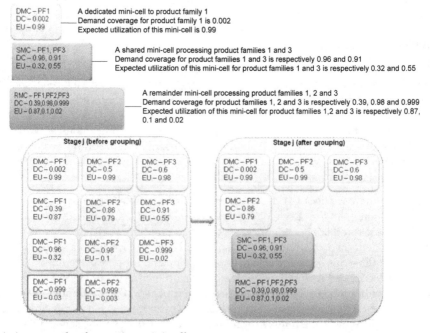

Fig. 4. An example of grouping mini-cells

3.3 Locating facility and allocating mini-cells

A capacitated plant location mathematical model is built to allocate mini-cells to the candidate plants. The objective of this model is to minimize the total cost including production cost, investment cost, and transportation cost as given in Equation 9. Candidate plants are located in different areas with limited capacities, and various production and investment costs. Transportation costs include the costs of transporting products between two consecutive manufacturing stages, and also the costs of shipping products to the market. A single market is assumed. Equation 10 guarantees that the capacity allocated to a plant does not exceed the maximum available capacity. A mini-cell can be assigned to only one facility as presented in Equation 11. Equations 12 and 13 maintain transportation balance, in other words, the quantity of products shipped into a plant should match product quantity shipped out of this plant. Equation 14 enforces investment cost of opening a manufacturing stage in a plant. Parameter EQ_{ik} is roughly estimated as given in Equation 8, where EU_{ik} is the expected utilization of product family in in mini-cell k, and PT_{ij} is the processing time (in minutes) of bottleneck operation in stage j for family i.

$$Q_{ik} = 40 \times 60 \times EU_{ik} / PT_{ij} \tag{8}$$

Indices:

i Product family index
j Manufacturing stage index
k Mini-cell index
m Plant index

Parameters:

I Number of product families
J Number of manufacturing stages
K Number of mini-cells required
M Number of potential plants
NM_k Number of machines/workforce in mini-cell k
U_{jk} 1, if mini-cell k performs operations in manufacturing stage j; 0, otherwise.
EQ_{ik} Estimated quantity of product family i produced in mini-cell k
LOI_{ij} Previous stage index of stage j for product family i. 0 implies stage j is the first stage for family i or family i does not require manufacturing stage j
$MAXC_{jm}$ Available number of mini-cells for stage j in plant m
IC_{jm} Weekly equivalent investment cost for stage j in plant m
PC_{jm} Production cost for stage j in plant m (\$/40hour)
UTC Unit transportation cost (\$/mile/unit)
D_{mn} Distance from plant n to plant m
DM_m Distance from plant m to market
M Big value

Decision variables:

X_{km} 1, if mini-cell k is allocated to plant m; 0, otherwise

W_{jm} 1, if stage j is opened in plant m; 0, otherwise

TQ_{ijmn} Transportation quantity of family i from plant n to plant m from stage j-1 to stage j

Objective Function:

$$\min \quad Z = \sum_{j=1}^{J} \sum_{m=1}^{M} (PC_{jm} \times \sum_{k=1}^{K} (NM_k \times U_{jk} \times X_{km}) + IC_{jm} \times W_{jm})$$

$$+ \sum_{i=1}^{I} \sum_{j=1}^{J} \sum_{n=1}^{M} \sum_{m=1}^{M} (TQ_{ijmn} \times D_{mn} \times UTC) \tag{9}$$

$$+ \sum_{m=1}^{M} \sum_{i=1}^{I} \sum_{k=1}^{K} (X_{km} \times U_{Jk} \times EQ_{ik} \times DM_m \times UTC)$$

Subject to:

$$\sum_{k=1}^{K} (X_{km} \times U_{jk}) \le MAXC_{jm} \quad \text{for } j = 1, \cdots, J \ \& \ m = 1, \cdots, M \tag{10}$$

$$\sum_{m=1}^{M} X_{km} = 1 \quad \text{for } k = 1, \cdots, K \tag{11}$$

$$\left. \sum_{n=1}^{M} TQ_{ijmn} = 0 \right|_{LOI_{ij}=0}$$
$$\left. \sum_{n=1}^{M} TQ_{ijmn} = \sum_{k=1}^{K} (X_{km} \times U_{jk} \times EQ_{ik}) \right|_{LOI_{ij} \ne 0} \quad \text{for } i = 1, \cdots, I \ \& \ j = 1, \cdots, J \ \& \ m = 1, \cdots, M \tag{12}$$

$$\left. \sum_{m=1}^{M} TQ_{ijmn} = 0 \right|_{LOI_{ij}=0}$$
$$\left. \sum_{m=1}^{M} TQ_{ijmn} = \sum_{k=1}^{K} (X_{km} \times U_{LOI_{ij}k} \times EQ_{ik}) \right|_{LOI_{ij} \ne 0} \quad \text{for } i = 1, \cdots, I \ \& \ j = 1, \cdots, J \ \& \ n = 1, \cdots, M \tag{13}$$

$$M \times W_{jm} \ge \sum_{k=1}^{K} (X_{km} \times U_{jk}) \quad \text{for } j = 1, \cdots, J \ \& \ m = 1, \cdots, M \tag{14}$$

An example of allocation process is demonstrated in Figure 5. The results indicate that four mini-cells including three dedicated mini-cells and one reminder mini-cell are allocated to plant 3. One dedicated mini-cell and one shared mini-cell are to plant 4. Plant 1 is not able to implement any operation of manufacturing stage j, thus, there is no mini-cell allocated to this plant. Plant 2 is not chosen either based on various factors such as capacity, production cost, and distances.

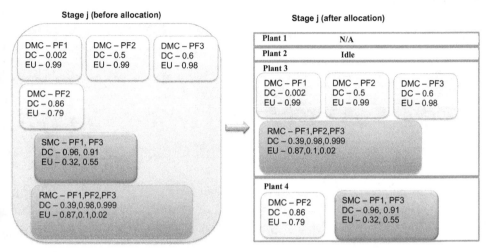

Fig. 5. An example of facility location and mini-cell allocation

3.4 Multi-period production planning

The capacitated plant location model selects plants and determines how many mini-cells should be built in each plant for each manufacturing stage. To continue studying on supply chain operational decisions, a production planning model is proposed to help decide production quantity in each mini-cell for manufacturing stage j in each week. The objective is to minimize demand shortage as given in Equation 15. Equation 16 computes demand shortage for each product family. Equation 17 guarantees that the product family is not assigned to a mini-cell which does not handle that family. Each cell only functions up to 40 hours per week as given in Equation 18. The boundary of decision variable is defined in Equation 19.

Indices:

i Product family index
k Mini-cell index

Parameters:

I Number of product families
K Number of mini-cells required
D_i Demand of product family i for current week
CU_{ik} Expected utilization of family i in mini-cell k
PT_i Processing time (in minutes) of bottleneck operation of family i
M Big value

Decision variables:

Q_{ik} Quantity of product family i produced in mini-cell k
DS_i Demand shortage of product family i

Objective Function:

$$\min \quad Z = \sum_{i=1}^{I} DS_i \tag{15}$$

Subject to:

$$DS_i = \max\left\{0, D_i - \sum_{k=1}^{K} Q_{ik}\right\} \text{ for } i = 1, \cdots, I \tag{16}$$

$$Q_{ik} \le M \times CU_{ik} \text{ for } i = 1, \cdots, I \ \& \ k = 1, \cdots, K \tag{17}$$

$$\sum_{i=1}^{I} (PT_{ik} \times Q_{ik}) \le 40 \times 60 \text{ for } k = 1, \cdots, K \tag{18}$$

$$Q_{ik} \ge 0 \tag{19}$$

An example of production plan giving detailed production quantity of each mini-cell for manufacturing stage *j* in week *n* is shown in Figure 6. For example, in plant 3, a dedicated mini-cell to product family 1 needs to produce 35 units of product family 1 in week *n*.

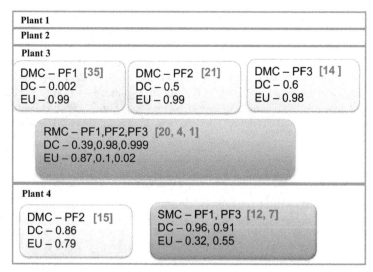

Fig. 6. An example of weekly production planning result

4. The system studied and the preliminary results

An example of supply chain involving a three-stage manufacturing system is studied in this section. This system was originally inspired from a global jewelry manufacturing company. The candidate production plants are mainly located in Caribbean, East Asia, and South East Asia. The jewelry products change along the fashion trend therefore resulting in highly fluctuating demand. It is very challenging to manage capacity to satisfy demand without reserving too much capacity.

The system studied consists of 12 product families with normally distributed weekly demand. The standard deviation is 25% of the average demand. Up to eight operations are required to manufacture a product, and they are grouped into three operation groups since some production facilities are not able to perform some of the operations (e.g. plating operation). All products go through three stages in the same order. In the jewelry manufacturing process, different parts within one product family always require the same operations and processing times are very close, thus, Table 4 shows the processing times and weekly demand for each product family instead of each part. This chapter focuses on supply chain design of manufacturing tier, thus, only one market is assumed.

	Stage	Manufacturing Operations	PF1	PF2	PF3	PF4	PF5	PF6	PF7	PF8	PF9	PF10	PF11	PF12
Processing Times (Minutes)	1	Findings (F)		10.6			10.8							11.1
		Casting (C)	10.2		9.7	10.9		13.3	12.6	10.1	10.8	12.5	9.9	
		Tumbling(T)	10.8	11	9.9	12.6	11.2	13.9	13.5	12.3	11.2	13	13.4	12.2
	2	Plating (PL)	9.6	10.9	11.2	9.3	10.1	10.5	9.2	9.2	8.5	12	12.5	11.7
	3	Stone Setting(SS)	9.1	4.6			6.2	4.1				9.3		
		Enameling(E)				5.8	7.5		7.1	7.2		5.6		8.5
		Oven (O)				8.7	6.0		4.9	9.2		8.9		5.5
		Packaging(PA)	12.7	12.8	11.2	11.2	11.4	10.9	10.8	12.7	11.5	12.1	12.3	12.5
Weekly Demand (Units)		mean	1890	795	478	2127	722	520	2582	981	1773	1150	966	474
		SD	472	199	120	532	180	130	646	245	443	288	242	118

Table 4. Processing times and demand

Experimentation of supply chain design on the studied system is implemented by using the proposed four-phase approach. For a three-stage manufacturing system with 12 product families, the mini-cell requirements and expected mini-cell utilizations are computed. Cells consist of multiple machines and equipment. Each machine/equipment requires an operator as well. Table 5 summarizes the number of mini-cells required for each product family as well as the total number of machines/workforce for each stage. The results indicate that a total of

381 mini-cells are required. These 381 mini-cells require a total of 708 machines/equipment and also 708 operators. The number of machines/equipment and operators needed for stages 1, 2, and 3 are computed as 272, 113 and 323, respectively. Due to the space limit, Table 6 only shows partial utilization results for stage 1 for product families which require no more than 6 mini-cells. It is important to notice that many cells are rarely utilized as shown in Table 6 such as 4th, 6th, 6th, and 5th mini-cells for product families 3, 5, 6, and 12, respectively.

	PF	Stage 1	Stage 2	Stage 3	Total Number
	1	16	14	18	
	2	7	7	8	
	3	4	4	4	
	4	20	15	18	
	5	6	6	7	
Number of	6	6	5	5	381
Mini-Cells	7	26	18	21	
	8	9	7	10	
	9	15	12	16	
	10	12	11	11	
	11	10	9	9	
	12	5	5	5	
Number of Machines/Workforce		272	113	323	708

Table 5. Summary of capacity requirement results

Mini-cell	PF3	PF5	PF6	PF12
1st	0.995367	0.999409	0.999131	0.998108
2nd	0.792689	0.98242	0.969661	0.91366
3rd	0.180275	0.835848	0.736627	0.446252
4th	0.003406	0.441633	0.272882	0.050731
5th		0.101697	0.032402	0.000743
6th		0.008137	0.000961	

Table 6. Partial results of mini-cell utilization for stage 1

Heuristic procedure is applied to form dedicated, shared and remainder mini-cells. The results of mini-cell formation are summarized in Table 7. Demand coverage is set to be 0.96 for the heuristic procedure. In other words, product demand will be covered 96% of the time. It is observed that majority of mini-cells are dedicated mini-cells, thus, the operational complexity is limited. By grouping product families into shared and remainder cells, mini-cell requirement is reduced from 381 to 210. Total number of machines/workforce is reduced from 708 to 414.

		Stage 1	Stage 2	Stage 3	Total Number
Number of Mini-Cells	Dedicated	50	40	49	210
	Shared	18	10	17	
	Remainder	8	11	7	
Number of Machines/Workforce		158	61	195	414

Table 7. Summary of mini-cell formation results

There are seven potential production facilities performing operations for different manufacturing stages as shown in Table 8, where 0 implies that the plant doesn't perform operations in this manufacturing stage. For example, plant 4 only performs the plating operation. Distance matrix is given in Table 9. Production costs vary from 32 to 850 representing huge gap of labor costs between developed areas and developing areas.

The math model is solved by ILOG OPL software. The allocation of capacity to production facilities is determined as shown in Figure 7. Plants 1 and 3 are not chosen to perform any operation due to their high production costs. However, production costs are not the only criteria of making decisions. For example, for manufacturing stage 3, production costs in plants 5 and 6 are very low compared with plant 2. But they are not chosen considering the high transportation costs.

	Stage 1	Stage 2	Stage 3
Plant 1	50/800/30000	40/835/30000	50/840/60000
Plant 2	30/600/5000	0	30/680/5000
Plant 3	60/650/10000	50/850/50000	75/720/30000
Plant 4	0	50/520/5000	0
Plant 5	0	0	50/32/25000
Plant 6	50/76/5000	0	50/61/5000
Plant 7	0	30/52/5000	50/72/5000

Table 8. Capacity (in 40 hours)/production cost ($40 hours) /investment cost

Plant	1	2	3	4	5	6	7
1	0	30	50	300	9000	8300	8500
2	30	0	40	260	8960	8280	8530
3	50	40	0	280	9030	8300	8500
4	300	260	280	0	10200	9300	9200
5	9000	8960	9030	10200	0	500	540
6	8300	8280	8300	9300	500	0	70
7	8500	8530	8500	9200	540	70	0
Market	300	350	310	300	7500	5600	5500

Table 9. Distance Matrix

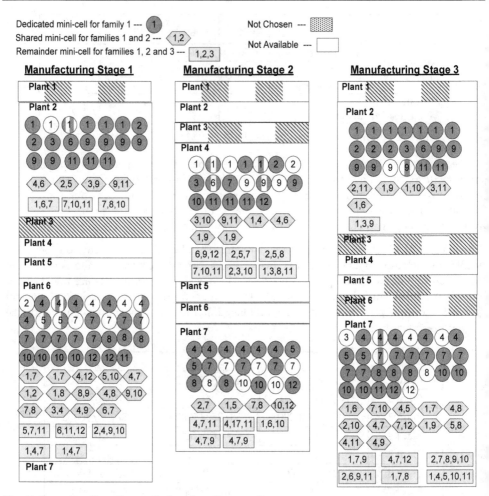

Fig. 7. Capacity allocation and plant location results

For multi-period production planning, the experimentation runs for a year (50 weeks). In each week, demand is randomly generated based on normal distribution. The proposed production model under such demand is solved by using OPL software, the demand shortage for each stage is recorded, and next, it is compared to the results of classical cellular system.

5. Comparison with classical mini-cellular system

In this section, performance of the proposed layered mini-cellular system is compared to that of a classical cellular system. Machine/workforce requirements and demand shortage are used as performance measures to evaluate these two systems. The proposed capacitated plant location model is also capable of solving resource management problems when classical cellular design is adopted in the production facility. However, the model parameters such as NM_k, U_{jk}, and EQ_{ik} are computed differently, since each mini-cell is dedicated to a single product family in the classical cellular system. The number of mini-cells required by product family i in stage j is computed in Equation 20, where μ_i is demand mean of family i, and PT_{ij} is the processing time (in minutes) of bottleneck operation for family i in stage j. RCU is the reserved cell utilization in order to handle high demand situation. Obviously, the value of RCU affects both performance measures: machine/workforce requirements and demand shortage. Reserving too little capacity leads to high demand shortage; while reserving too much capacity results in redundant machine/workforce therefore increasing production costs. In this section, the preliminary experimentation is carried out to illustrate the procedure, thus, RCU is set to be 10%. In the future, experimentation with various levels of RCU will be implemented and results will be studied.

$$NC_{ij} = \mu_i \times PT_{ij} / (40 \times 60 \times (1 - RCU)) \qquad (20)$$

The capacity requirement is computed for each product family at each stage, and the results are summarized in Table 10. It is observed that classical system requires 176, 73, and 204 machines/workforce for manufacturing stages 1, 2, and 3, respectively; while layered system only requires 158, 61, and 195 machines/workforce.

	PF	Stage 1	Stage 2	Stage 3	Total Number
	1	9	12	10	
	2	5	5	5	
	3	3	3	3	
	4	10	12	13	
	5	4	4	4	
Number of Mini-Cells	6	3	3	4	245
	7	11	13	17	
	8	5	6	6	
	9	7	10	10	
	10	7	7	7	
	11	6	6	6	
	12	3	3	3	
Number of Machines/Workforce		176	73	204	453

Table 10. Summary of capacity requirement results of classical design

The performance of the proposed layered mini-cellular system in terms of handling fluctuating demand is investigated in this section. Fifty demand sets are randomly generated based on normal distribution given in Table 4. The service level for each period is computed based on the demand supplied from the facility network divided by total demand. The results are obtained by using both the layered design and classical cellular design with 10% reserved cell utilization (as shown in Figure 8). For manufacturing stage 1, it can be observed that layered mini-cellular system leads to a high service level (>90%) most of the time. There are only six out of 50 periods when the service level is below 90%. Under most conditions, layered system leads to a higher service level. There are only seven exceptions where classical design leads to a higher service level. The similar pattern could be also observed for manufacturing stages 2 and 3. It is clearly observed that, compared to the classical system, the layered system model requires less number of mini-cells and machines/workforce while still dealing with high demand fluctuation more effectively as evidenced by higher service levels.

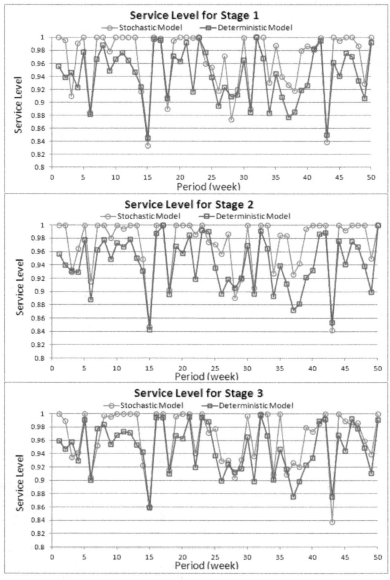

Fig. 8. Demand shortage for each manufacturing stage

6. Conclusion

This chapter studies the design of a supply chain involving multi-stage manufacturing operations with probabilistic product demand. Three levels of supply chain issues are first discussed. In the strategic level, a 'transnational vertical integration' market and production location strategy is taken, where the multi-stage manufacturing system is across various geographical locations and finished goods are sold in North American market. The study then mainly focuses on determining how much capacity should be allocated to which production facility for each manufacturing stage.

Manufacturing configuration in each individual facility is also taken into account in this chapter. The chapter integrates manufacturing system design with supply chain design by proposing a layered mini-cellular system. Each mini-cell is assumed to operate one manufacturing stage with maximum 40 hours weekly capacity. In the classical cellular system, a cell is dedicated to one product family. A layered system not only consists of dedicated mini-cells but also shared and remainder mini-cells. Mini-cell requirements and utilization are first estimated by using probability equations. Mini-cells are then grouped based on operation similarities among product families, and eventually, dedicated, shared and remainder mini-cells are formed. The latter two types of mini-cells deal with more than one product family so that resources are shared and demand fluctuation could be neutralized to a certain level.

A capacitated plant location math model is proposed to form supply chain network as well as allocate mini-cells to each facility for each manufacturing stage. Both mini-cell components and transportation costs are taken into account this model. Next, a capacity planning model determines detailed production quantity based on a specific weekly demand.

Experimentation is conducted and the results indicate the selection of production facilities and allocation of capacity. The performance of layered system is compared with the results of the classical cellular manufacturing system. It is important to notice that despite the lower number of machines/workforce was required by the layered system; the layered system satisfies demand better compared to the classical system. The results indicate that this study provides a complementary analytical model that explores the efficient way to locate and allocate inbound resources so that a certain level of supply chain efficiency and responsiveness could be achieved.

7. References

Chopra, S., & Meindl, P. (2007). Supply *Chain Management-Strategy: Planning, & Operation* (third ed.). New Jersey: Prentice-Hall, Inc., ISBN 978-013-6080-40-4, New Jersey

Cosner, J. D. (2008). Technology, Location, Price, and System Design Decisions for a Global Manufacturing Company. Master's Thesis, Industrial and Systems Engineering, Ohio University, Available from
http://etd.ohiolink.edu/view.cgi/Cosner%20Jeremy%20D.pdf?ohiou1225542925

Dicken, P. (1992). *Global Shift: The Internationalization of Economic Activity* (second ed.). London: Paul Chapman Publishing, ISBN 978-185-3961-42-7, London

Lee, H. L. (2000). Creating Value through Supply Chain Integration, *Supply Chain Management Review*, Vol.6, pp. 30-37, ISSN 1521-9747

Lee, H. L. (2005). The Triple-A Supply Chain, *Harvard Business Review*, Vol. 683, No. 1, pp. 112-113, ISSN 0017-8012

Rao, P. P., & Monhanty, R. P. (2003). Impact of cellular manufacturing on supply chain management: exploration of interrelationships between design issues. *International Journal of Manufacturing Technology Management*, Vol. 5, pp. 507-520, ISSN 1368-2148

Schaller, J. (2008). Incorporating cellular manufacturing into supply chain design, *International Journal of Production Research*, Vol. 46, No. 17, pp. 4925-4945, ISSN: 0020-7543

Suer, G. A., Huang, J., & Maddisetty, S. (2010). Design of dedicated, shared and remainder cells in a probabilistic demand environment. *International Journal of Production Research*, Vol. 48, No. 19, pp. 5613-5646, ISSN: 0020-7543

9

Facility Layout

Xiaohong Suo
China Institute of Industrial Relations, Beijing,
P.R.China

1. Introduction

The facility layout problem is concerned with finding the most efficient non-overlapping arrangement of n indivisible departments with unequal area requirements within a facility. Generally, about 20%-50% of the total operating expenses in manufacturing are attributed to material handling costs. Effective facility layout could reduce these costs by 10%-30% annually. Moreover, good facility planning could also improve the material handling efficiency, reduce the throughput time, decrease the space utilization area of manufacturing system, etc. So, the facility layout affects the total performance of manufacturing system, such as, material flow, information flow, productivity, etc.

Facility layout, being a significant contributor to manufacturing performance, has been studied many times over the past decades. Raman et al. showed that facility layout has a direct impact on operational performance, as measured by manufacturing lead time, throughput rate, and work-in-process (WIP).

2. Classification of facility layout problem

It is well known, facility layout problem is concerned with the allocation of activities to space such that a set of criteria are met and/or some objectives are optimized. There are numerous derivations for the facility layout problems in manufacturing systems, which have been investigated in Table 1. These derivations can be classified into six categories (product, process, equipment, production, manufacturing system and company). Any changes in items of these six categories can lead to facility layout problem. Once one item changes, other items will change correspondingly, e.g., the introduction of new products results in changes in process and equipment. Generally, combinations of items of these six categories are the derivations for the facility layout problem.

When the flows of materials between the departments are fixed during the planning horizon, facility layout problem is known as the static (single period) facility layout problem (SFLP). Researchers had paid more attentions to SFLP, and now SFLP has two new trends. With more fierce competitive in the global market, facility layout must react on the changes in designs, processes, quantities, scheduling, organizations, and management idea rapidly. Once these items change frequently, manufacturing systems must be reconfigurable and their structure must be modified as well. However, SFLP can hardly meet this demand. Company need to design a flexible layout which is able to modify and expand easily the original layout. Flexibility can be reached by modular devices, general-purpose devices and material handling devices. The trends of SFLP are shown in Fig.1. Under a volatile environment, SFLP need to add flexibility to meet the production requirement. Approaches to get flexibility for SFLP include to modify the SFLP and to increase the robustness of the SFLP. Gradually, SFLP develops these two approaches to the dynamic facility layout problem (DFLP) and robust layout, respectively.

Up to now, there are existing three basic types of layout problem, including SFLP, DFLP and robust layout problem. Research on the relationships among SFLP, DFLP and robust layout is important due to the impact of the types of layout problem on productivity, quality, flexibility, cost, etc. How to select the suitable type of layout problem is an urgent task. The classification procedure of facility layout problem is shown in Fig.2, where researchers choose the appropriate type of layout problem based on the judgment conditions. The judgment conditions include whether the material handling flows change over a long time or not, and whether it is easy for rearrangement or not when the production requirement changes drastically. If the material handling flows change over a long time, choose DFLP or robust layout; if not, choose SFLP. If rearrangement is easy when the production requirement change drastically, choose DFLP; if not, choose robust layout.

No. Classify	1	2	3	4	5
Product	Increase or decrease in the demand for a product	Addition or deletion of a product	Changes in the design of a product	Introduction of new products	
Process	Changes in the design of process	Replacement of characteristics of process	Installation of new processes		
Equipment	Installation of new equipment	Replacement of one or more pieces of equipment			
Production	Failure to meet schedules	High ratio of material handling time to production time	Excessive temporary storage space	Bottlenecks in production	Crowded conditions
Manufacturing systems	Conflict between productivity and flexibility in general manufacturing systems	Flexibility does not meet the demands of changes in product mixes of FMS.			
Company	Adoption of a new safety standard	Organizational changes within the company	A decision to build a new plant		

Table 1. Derivations for facility layout in manufacturing systems

Fig. 1. Trends of SFLP

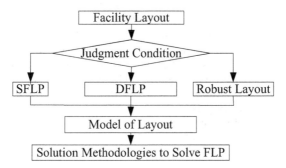

Fig. 2. Classification procedure of facility layout problems

2.1 Relationships among SFLP, DFLP and robust layout

Due to the impact of the locations of facility on material handling costs, throughput, and productivity of the facility, facility layout is an important module of manufacturing systems design. The FLP is the arrangement of departments within a facility with respect to some objective. The most common objective considered is the minimization of material handling cost. Material handling costs are determined based on the amounts of materials that flow between the departments and the distances between the locations of the departments. SFLP is appears when the flows of materials between departments are fixed during the planning horizon, which can be formulated as a quadratic assignment problem (QAP). So, SFLP is used under the static environment. When the flows of material between departments change during the planning horizon, this problem is known as the dynamic (multiple-period) facility layout problem (DFLP) [1]. Therefore, DFLP is widely used when the condition is changeable and the future demand of product can be forecasted. A robust layout is one that is good for a wide variety of demand scenarios even though it may not be optimal under any specific ones [2]. A robust layout procedure considers minimizing the total expected material handling costs over a specific planning horizon. Robust layout is selected when the demand is stochastic and the re-layout is prohibited.

2.1.1 SFLP vs. DFLP

SFLP Converting to DFLP.

Fiercer competition of the world makes SFLP covert to DFLP. Under today's changeable market situation, demand is changed irregularly from one production period to another. Generally, 40% of a company's sales come from new products, *i.e.* products that have only recently been introduced [3]. When these changes frequently occur and the location of an existing facility is a decision variable, SFLP convert to DFLP. The procedure of SFLP converting to DFLP is given in Fig. 3. The changes in product, process, equipment, etc. can bring on the facility layout problem. If the material flows are consumed to be constant, SFLP is sufficient. However, this assume are contradiction with the practice production. In order to correct the deficiency, SFLP will be converted to DFLP.

During the process of SFLP converting to DFLP, rearrangement costs arise. The changes in locations of facility can reduce the material flows between department pairs during a planning horizon. Meanwhile, rearranging the locations of facility will result in some shifting (rearrangement) costs depending on the departments involved in this shift. The procedure for rearrangement costs is illustrated in Fig. 4. Generally, when the products change often and the facility location is static, the material flows are increased drastically. In order to reduce the material flows, the facilities are shifted to different location which will result in the rearrangement (shifting) costs. The DFLP is based on the anticipated changes in flow that will occur in the future. Moreover, the future can be divided into any number of time periods, and a period may be defined in months, quarters, or years. In addition, different periods can be of different lengths [4].

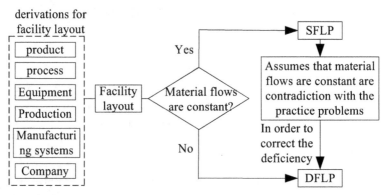

Fig. 3. Procedure of SFLP converting to DFLP

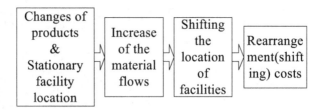

Fig. 4. Procedure for rearrangement costs in DFLP

SFLP for each period and rearrangement are the two parts of DFLP. For DFLP, it is assumed that the flow data during each period remains constant, respectively. Therefore, the facility layout during one period in the planning horizon can be obtained by solving the SFLP for each period. However, the flow data in whole planning horizon are changeable in DFLP. There exist rearrangement costs between the layouts for each pair of adjacent periods. That is to say, DFLP is composed of a series of SFLP if the rearrangement costs can be neglected. Therefore, DFLP involves selecting a SFLP for the first period and then deciding whether to change to a different SFLP in the next period. The dynamic layout shows flow dominance.

For each period, some departments have higher material handling inflows than the others. During the adjacent periods, if the higher material handling inflows do not change, the same SFLP will be used in these two periods. If these flow dominant departments change during the adjacent periods, changing to a different SFLP will occur in the following period.

For DFLP, the cycle of rearrangement depends on the rearrangement costs. If the rearrangement costs are relatively low, the layout configuration would tend to change more often to retain material handling efficiency. The reverse is also true for high rearrangement costs. The structural diagram of DFLP is given in Fig.5. Table 2 gives the comparison of main characteristics for SFLP and DFLP.

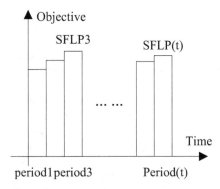

Fig. 5. Structural diagram of DFLP

	Period number	Rearrangement	Optimize objectives	Method for generating static layout of each period
SFLP	One period	No rearrangement	Minimize the material handling costs	The best static layout
DFLP	Multiple periods	Rearrangement costs	Minimize the material handling costs and rearrangement costs	The best static layout, or the random layout, or mixed layout*

*Mixed layout: combination of random layouts and the best layouts, each contributing half of the layout selected.

Table 2. Comparison between SFLP and DFLP

Objective Function of DFLP.

The objective function of a DFLP is generally defined as the minimization of the total costs, material handling costs for a series of SFLP plus rearrangement costs between periods. In each period, material handling costs among departments are calculated by the product of the probability of a flow matrix occurring in that period, the associated flows, and the distances. The formulation of the DFLP is given below.

$$\text{Min DFLP} = \sum_{t=1}^{T} \text{SFLP}(t) + \sum_{t=1}^{T-1} A_{t(t+1)} \tag{1}$$

where t is the number of periods in DFLP; $A_{t(t+1)}$ is the rearrangement costs between each pair of adjacent periods; and T is the total number of periods in the whole period horizon. Rearrangement costs are incurred when moving machines (or departments) from one location to another in order to minimize material handling costs in consecutive periods. Rearranging the layout will result in some shifting costs depending on the departments of the layouts. Rearranging includes the changes in location and orientation [5]. Each of the two aspects for facility will impact the total rearrangement costs. When the location and orientation of a facility in the new layout are the same as in the existing layout, no rearrangement cost is incurred for that facility. Otherwise, if the location or orientation of a facility has changed, then the specific facility rearrangement costs will be added to the objective function.

There are many approaches to calculate rearrangement costs for DFLP. Furthermore, the rearrangement of departments may lead to production loss, and it may also require specialized labor and equipment. Therefore, rearrangement costs consist of labor cost, equipment cost, out-of-pocket moving expenses and the cost of operational disruptions. Generally, the rearrangement (shifting) costs may be viewed as fixed costs, or a linear function of the distance between the various locations, or the linear function of square-feet being rearranged, or variable costs associated with moving a particular facility in a given period, or the accumulation of fixed costs due to changing in facility configuration, interrupting or disrupting production, using personnel and equipment to move the facility, or any combination of the above [5,6]. One special application of the DFLP is the rearrangement of existing facilities. In a rearrangement problem, the first period is the current SFLP and subsequent periods are the revised layouts.

Computation Complexity of DFLP.

As the SFLP, the DFLP is also the computationally intractable problem. In other words, the number of possible solutions or layout plans is $N!$ for a SFLP instance with N departments, while $(N!)^T$ for a DFLP instance with N departments and T periods. For this reason, with the same computer's configuration, only small problems can be optimally solved in reasonable computation time for DFLP while large and media problems can be solved for SFLP. The numbers of possible solutions and their methodologies for SFLP and DFLP are listed in Table 3.

For instance, even for a six-department, five-period problem, $(N!)^T$ is 1.93×1014 combinations. Thus for large problems obtaining optimal solutions is not nearly possible. So in practices, for small problem, $n=N!$; for large problem, $n<<N!$, where n is the number of static layout during each period in DFLP. n in each period depends on the capability of the software and hardware used to solve the DFLP. The more power these are, the larger n can be selected. Logically, larger n should lead to better solutions.

	Number of solutions	Size of problems	Methodologies
SFLP	$N!$	Small	Exact algorithm
		Media and large	Heuristic or meta-heuristic algorithm
DFLP	$(N!)^T$	Small and media	heuristic or meta-heuristic algorithm

Table 3. Numbers of solutions and methodologies for SFLP and DFLP

DFLP Degenerates into SFLP.

The SFLP has one period and no rearrangements, so it is just a special case of the DFLP. Under some conditions, DFLP may degenerate into SFLP. In those environments where material handling flows do not frequently change over a long time, SFLP analysis would be sufficient. When the rearrangement costs are negligible, dynamic layout analysis is not necessary. In the other cases, if the rearrangement costs are prohibitive, such as in the case of very heavy machinery, the same layout is used for the total planning horizon. In this situation, DFLP is also not necessary.

2.1.2 DFLP vs. robust layout

Robustness is defined as the frequency that a layout falls within a pre-specified percentage of the optimal solution for different sets of production scenarios [7]. For a robust layout, it is good for a wide variety of demand scenarios even though it may not be optimal under any specific demand scenarios. The objective of a robust layout is trying to minimize the total material handling costs over the specific planning horizon. Fig.6 illustrates the robust layout design framework [6].

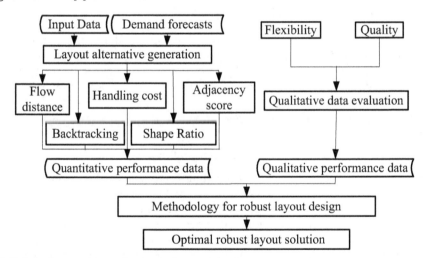

Fig. 6. Robust layout design framework

To some extent, robust layout is similar to SFLP. However, there is a difference between them: robust layout is researched under dynamic environment, and SFLP is assumed that the environment is static. Under the volatile environment, if the total planning horizon is divided into a lot of periods, DFLP is considered to be composed of a series of SFLP. Under some condition, if the number of periods is small over the planning horizon, *i.e.* each period has long time, DFLP is considered to be composed of a series of robust layout based on the definition of robust layout.

For DFLP, a set of planning time are referred to as a consecutive period with a layout rearrangement occurring only at the beginning of each period. The number and length of these periods are determined based on the trade-off between material handling costs and facility rearrangement costs. If rearrangement costs are larger than material handling costs, the number of these periods will be small and length will be long. The reverse is also true. Thus, the aim of DFLP is to modify the layout at the beginning of each period, but not to change the layout within these periods. In Contrast to DFLP, when rearrangement costs are extremely high, a pure robust layout will be selected, and it has the period equal to a total planning horizon. DFLP and robust layout are compared in Table 4. Robust layout will have different demand levels in the total planning horizon and will choose the demand level to minimize the material handling costs, adjacency scores and backtracking costs.

	Selective conditions	Period number	Function for material handling costs
DFLP	if production requirements change drastically and the rearrangement is easy	Multiple periods	Between the lower bound and upper bound on the expected material handling costs
Robust layout	if machine rearrangement costs are high	One period equal to the total planning horizon	Provide an upper bound on the expected material handling costs

Table 4. Comparison between DFLP and robust layout

As described in the last column of Table 4, robust layout provides an upper bound on the expected material handling costs while creating a new layout for each period provides a lower bound [8]. But in practical problems, creating a new layout for each period is unrealized. Therefore DFLP, which material handling costs are between the lower bound and upper bound, is necessary.

2.2 Discussions

The comparison among SFLP, DFLP and robust layout is presented in Table 5.

The fifth column of Table 5 refers the converting conditions of SFLP, DFLP and robust layout with each other. When the flows between department pairs are changeable, the SFLP will convert to DFLP or robust layout. When rearrangement costs are negligible, DFLP will convert to SFLP. However, when rearrangement costs are extremely high, DFLP will convert to robust layout. Under stochastic demand and when forecasting the future demand of products is difficult, robust layout will convert to DFLP. Since studies show that material handling costs make up 20-50% of the total operating costs and 15-70% of the total costs of manufacturing a product [9], the most common objective considered are the minimization of material handling costs. For DFLP, its objective function involves material handling costs and rearrangement costs. However, the objective of robust layout includes the total material handling costs for all product mix in total planning horizon.

The connection and difference between each pair of the types of layouts are described in Table 6.

Since assuming that the material handling flows are constant for SFLP, uncertainly of future production requirements are relatively low. DFLP is suitable for dynamic environment, so its uncertainty of future production requirements is high. Although SFLP and robust layout both have one period and no rearrangement, the application scopes are different-one for static environment and the other for dynamic environment. DFLP and robust layout are both used under the dynamic and changeable environment, but DFLP considers the rearrangement and robust layout does not consider for only using one facility layout solution during the whole planning horizon.

	Application scope	Advantages	Disadvantages	Converting condition	Objective function
SFLP	Flows between pairs of departments do not change over a long time	Widely used; low computational complexity; modeling easily	Flows between departments are assumed to be constant over time	Flows between department pairs are changeable	$\min \sum\limits_{i=1}^{n} \sum\limits_{j=1}^{n} c_{ij} f_{ij} d_{ij}$ $i, j = 1, 2, ..., n$
DFLP	The relative material flows between departments change over time	Moving expenses and operational disruption	High computational complexity	Rearrange-ment costs are either negligible or extremely high	$\min(\sum\limits_{t=1}^{P} \sum\limits_{i=1}^{n} \sum\limits_{j=1}^{n} \sum\limits_{k=1}^{n} \sum\limits_{l=1}^{n} f_{tij} d_{tij} X_{t}$ $+ \sum\limits_{i=1}^{t-1} A_{t(t+1)})$
Robust Layout	Under stochastic demand and the re-layout is prohibited	Offer low mean and variance in distance traveled; lower material handling costs over large changes in product demand	It is not the best optimal solution for a special product.	Forecasting the future demand of products is difficult	$\min \sum\limits_{m=1}^{M} \sum\limits_{i=1}^{n} \sum\limits_{j=1}^{n} C_{mij} f_{mij} d_{m}$ $i, j=1,2,...,n; \ m \text{ product mix}$

Table 5. Comparison among SFLP, DFLP and robust layout

	SFLP	DFLP	SFLP	Robust layout	DFLP	Robust layout
Connection	SFLP is the base and the special case of DFLP		One period and no rearrangement		In the dynamic and changeable environment	
Difference	Uncertainty of future production requirements		Static environ-ment	Dynamic environ-ment	Multiple periods and rearrangement	One period and no rearrangement
	Low	High				

Table 6. Connection and difference between each pair of the types of layouts

2.3 Summaries

In this part, the relationship among SFLP, DFLP and robust layout are researched. The characteristics of SFLP, DFLP are analyzed first. Then DFLP and robust layout are compared. The research results are given as followings:

1.   The application scope of DFLP is different from SFLP and robust layout.

2. DFLP can convert to SFLP or robust layout in some conditions.
3. Among SFLP, DFLP and robust layout, the majority of practical problems will be classified to DFLP.
4. SFLP is the base of DFLP and robust layout.
5. SFLP and robust layout both are the special case of DFLP.

3. Objectives and constraints of FLP

The facility layout problem, block layout, considers the assignment of facilities to locations so that the quantitative (qualitative) objective of the problem is minimized (maximized) under various constrains.

3.1 Objectives of FLP

Traditionally, there are two basic types of objectives for FLP [10]. The first one is the quantitative (distance-based) objective aiming at minimizing the total material handling cost between departments based on a distance function. The distance-based objective, considers all distance pairs, but due to department areas, inter-department distances may be misleading. To help relieve this concern, distances have been measured in a variety of ways: from department centroid-to-centroid, expected distance, distance from department boundaries, distance along the material handling network, etc. Even so, since the choice between general-purpose and special-purpose material handling devices may depend on whether or not departments are adjacent, the same inter-department distance may not have the same material handling cost.

In general, the actual cost to move a unit load of material between two departments will be the sum of a fixed cost and a variable cost. The fixed cost is dependent on the waiting time to obtain the appropriate material handling method, the time to pickup and deposit the unit load, and possibly some charge for the initial purchase cost of the material handling method. The variable material handling cost is dependent on the distance the unit load travels. In single-floor facilities this has a near-linear relation, while in multi-floor facilities it is a non-linear relationship.

The second one is the qualitative (adjacency-based) goal, aimed at maximizing the closeness relationship scores between departments based on the placement of departments that utilize common materials, personnel, or utilities adjacent to one another, while separating departments for reasons of safety, noise, or cleanliness. The adjacency-based objective, if interpreted from a material handling cost perspective, is based on the assumption that the material handling costs between two departments are reduced significantly when the two departments are adjacent. The adjacency-based objective appears to assume that fixed costs dominate the total costs (to the extent we can ignore the variable material handling costs) and that more efficient (and less costly) material handling methods may be used when departments are adjacent. On the other hand, the distance-based objective models the variable material handling costs and ignores the fixed costs.

Over the years, extensive research has been conducted on FLP. Yet, most of the research conducted in this field has concerned a single objective, either qualitative or quantitative goodness of the layout. In general, minimization of the total material handling costs is often used as the optimization criterion in FLP. However, closeness, hazardous movement or safety, and similar criteria are also important in FLP. Inherently, real-life layout problems are multi-objective by nature and the FLP must consider both quantitative and qualitative objectives simultaneously. Consequently, FLP falls into the category of multi-objective optimization problem.

3.2 Constraints of FLP

Facility layout plays a crucial role in determining the throughout time of a manufacturing process. The objective of the facility layout problem in manufacturing environment is the arrangement of facilities on a floor shop [11], subject to the following constraints:

1. to reduce the flows among all facilities;
2. to have a regular flow of the parts and products not permitting bottleneck in the production;
3. to rationalize the space occupied by the facilities;
4. to permit flexibility considering that with the technological progress and the new demands in the market, facilities could be added or changed.
5. to locate in a specified location.

Facilities are including machines, departments, storage equipments, factory, material-handling systems, commerce and warehouse. In the manufacturing system it may be distinguished machines, material handling systems and storage equipments.

4. Mathematic formulation of facility layout problem

For the past decades researchers have been working on facility layout problem while considering various aspects which vary with the nature of production demand, shape of the facilities, number of floors, and nature of material flow. Despite these variations, the process of obtaining optimal solutions involves two steps: modeling the facility layout problem, and developing a solution approach. Modeling helps clearly define the problem and consider the factors that are imperative in developing layouts.

The facility layout problem is one of the best-studied problems in the field of combinatorial optimization. A number of formulations have been developed for this problem. Models are categorized depending on their nature, assumptions and objectives. More particularly the FLP has been modeled as quadratic assignment problem (QAP), quadratic set covering problem (QSP), linear integer programming problem (LIP), mixed integer programming problem (MIP), and graph theoretic problem.

4.1 Quadratic assignment problem (QAP)

Koopmans and Beckman were the first to model the problem of locating plants with material flow between them as a quadratic assignment problem (QAP) in 1957 [12]. The name was so given because the objective function is a second-degree function of the variables and the constraints are linear functions of the variables. More specifically, it is an NP-hard problem and one of frequently used formulation to solve FLP.

Consider the FLP of allocating a set of facilities to a set of locations, with the objective to minimize the cost associated not only with the distance between locations but with the flow also. Each location can be assigned to only one facility, and each facility can be assigned to only one location. There is material flow between the different departments and cost (material handling) associated with the unit flow per unit distance. Thus, different layouts have different total material handling costs depending on the relative location of the facilities. F_{ik} is the flow between facilities i and k, and D_{jl} is the distance between location j and l. The FLP has been formulated as follows:

$$\sum_{\substack{i=1 \\ i \neq k}}^{n} \sum_{\substack{j=1 \\ j \neq l}}^{n} \sum_{k=1}^{n} \sum_{l=1}^{n} F_{ik} * D_{jl} * X_{ij} * X_{kl} \tag{2}$$

$$s.t. \quad \sum_{i=1}^{n} X_{ij} = 1 \qquad \forall \ j = 1,...,n \tag{3}$$

$$\sum_{j=1}^{n} X_{ij} = 1 \qquad \forall \ i = 1,...,n \tag{4}$$

$$X_{ij} \in \{0,1\} \qquad \forall \ i,j = 1,...,n \tag{5}$$

$X_{ij} = 1$ if facility i is assigned to location j and $X_{ij} = 0$ if facility is not assigned to location j, where n is the number of facilities. Equation (2) seeks to minimize the sum of flow multiplied by the distance for all pairs of facilities in a given layout. Equation (3) ensures that each location contains only one facility while equation (4) ensures that each facility is assigned to only one location.

4.2 Quadratic set covering problem (QSP)

Bazaraa formulated facility layout problem as a quadratic set covering model in 1975 [13]. In this formulation, the total area occupied by all facilities is divided into a number of blocks where each facility is assigned to exactly one location and each block is occupied by at most one facility. The distance between the locations is taken to be from centriods of the locations and the flow between facilities is minimized. The disadvantage of this approach is that the problem size increases as the total area occupied by all the facilities is divided into smaller blocks.

4.3 Linear integer programming problems (LIP)

Several integer programming formulations have been proposed for the facilities layout problem. Lawler was the first one to formulate the FLP as a linear integer programming model [14]. He proved that his model is equivalent to QAP. QAP has n^2 k_{ij} variables and $2n$ constraints while integer programming problem has n^4+2n+1 constrains and n^4 Y_{ijkl} while n is the number of locations, X_{ij} is the integer variable of facility i at location j, and Y_{ijkl} is the integer variable of facility i at location j in arrangement k of location l.

Assumption $y_{ijkl} = x_{ij} \cdot x_{kl}$

$$\text{Min}\sum_{i=1}^{n}\sum_{j=1}^{n}\sum_{k=1}^{n}\sum_{l=1}^{n}b_{ijkl}y_{ijkl} \qquad (6)$$

$$s.t. \quad \sum_{j=1}^{n}x_{ij}=1, \ i=1,2,...,n; \qquad (7)$$

$$\sum_{i=1}^{n}x_{ij}=1 \quad j=1,2,...,n; \qquad (8)$$

$$\sum_{i=1}^{n}\sum_{j=1}^{n}\sum_{k=1}^{n}\sum_{l=1}^{n}y_{ijkl}=n^2 \qquad (9)$$

$$x_{ij}+x_{kl}-2y_{ijkl}\geq 0 \ i,j,k,l=1,2,...,n \qquad (10)$$

$$x_{ij}\in\{0,1\} \quad i,j=1,2,...,n. \qquad (11)$$

$$y_{ijkl}\in\{0,1\} \ i,j,k,l=1,2,...,n \qquad (12)$$

$$b_{ijkl}=\begin{cases} f_{ik}c_{jl}+a_{ij} & i=k,j=l \\ f_{ik}c_{jl} & i\neq k \ or \ j\neq l \end{cases} \qquad (13)$$

4.4 Mixed integer programming problems (MIP)

Kaufman and Broeckx developed a linear mixed integer programming model in 1979 [15], which has the smallest number of variables and constraints among all integer programming formulations of the QAP. The equivalence between QAP and the mixed integer programming has been proposed through this model.

Assumption $w_{ij}=x_{ij}\sum_{k=1}^{n}\sum_{l=1}^{n}b_{ijkl}x_{kl}$, $e_{ij}=\sum_{k=1}^{n}\sum_{l=1}^{n}b_{ijkl}$

$$\text{Min}\sum_{i=1}^{n}\sum_{j=1}^{n}\sum_{k=1}^{n}\sum_{l=1}^{n}b_{ijkl}x_{ij}x_{kl}=\sum_{i=1}^{n}\sum_{j=1}^{n}x_{ij}\left(\sum_{k=1}^{n}\sum_{l=1}^{n}b_{ijkl}x_{kl}\right)=\sum_{i=1}^{n}\sum_{j=1}^{n}w_{ij} \qquad (14)$$

$$s.t \ \sum_{j=1}^{n}x_{ij}=1 \ i=1,2,...,n;, \qquad (15)$$

$$\sum_{i=1}^{n}x_{ij}=1 \quad j=1,2,...,n;, \qquad (16)$$

$$e_{ij}x_{ij}+\sum_{k=1}^{n}\sum_{l=1}^{n}b_{ijkl}x_{kl}-w_{ij}\leq e_{ij} \qquad (17)$$

$$w_{ij} \geq 0 \quad i,j = 1,2,...,n. \tag{18}$$

$$x_{ij} \in \{0,1\} \quad i,j = 1,2,...,n. \tag{19}$$

4.5 Graph theoretic formulations

In graph theoretic formulations it is assumed that the desirability of locating each pair of facilities adjacent to each other is known [16]. In this model a closeness rating indicating desirability of locating facility i adjacent to facility j is assumed. The model seeks to maximize the closeness rating of the facilities.

5. Solution methodologies for facility layout problem

Several researches have been done in the facility layout problem. The solution methodologies for FLP can be divided into exact algorithms, heuristics and meta-heuristic algorithms [17]. The exact methods such as the branch-and-bound and cutting plane algorithm have been successfully applied to FLP when the number of facilities is less than 16. However, when the number of facilities is larger than 16, FLP cannot be solved optimally in reasonable time. In order to obtain good (near optimal) solution in a reasonable computational time, heuristics were developed. Recently, meta-heuristic approaches such as simulated annealing (SA), genetic algorithms (GA), tabu search (TS), and colony optimization have been successfully applied to solve large FLP.

5.1 Exact algorithms

Exact algorithms are clever version of exhaustive search approach. Branch-and-bound and cutting plane algorithms are used to solve the FLP modeled as QAP optimally. These exact algorithms are complete in the sense that the existence of a feasible solution and then the optimal solution can be determined with certainty once such exact algorithm is successfully terminated. The main disadvantage of these exact algorithms is that they entail heavy computational requirements when applied even to small size problems.

5.1.1 Branch and bound algorithms

Branch and bound methods are used to find an optimum solution of quadratic assignment formulated FLP because QAP involves only binary variables. In branch and bound algorithms, the solution procedure proceed on the basis of stage by stage or parallel search of single assignment or pairs of assignments of facilities to locations. At each stage back tracking occurs, certain assignments are excluded and the forward search process is resumed.

Only optimal solutions up to a problem size of 16 are reported in literature. Beyond n=16 it becomes intractable for a computer to solve it and, consequently, even a powerful computer can not handle a large instance of the problem.

5.1.2 Cutting plane algorithms

Cutting plane methods are exact algorithms for integer programming problems. They have proven to be very useful computationally in the last few years, especially when combined with a branch and bound algorithm in a branch and cut framework. Cutting plane algorithms work by solving a sequence of linear programming relaxations of the integer programming problem. The relaxations are gradually improved to give better approximations to the integer programming problem, at least in the neighborhood of the optimal solution. For hard instances that can not be solved optimality, cutting plane algorithms can produce approximations to the optimal solution in moderate computation times, with guarantees on the distance to optimality.

5.2 Heuristic algorithms

In order to obtain good (near optimal) solution in a reasonable computational time, heuristic algorithms were developed [18]. A heuristic algorithm can be defined as a well-defined set of steps for quickly identifying good quality solutions. The quality of a solution is defined by an evaluation criterion, e.g., minimize material handling cost, and the solution must satisfy the problem constraints. Basically, heuristic algorithms for FLP can be classified into four classes: construction algorithms, improvement algorithms, hybrid algorithms and graph theoretic algorithms.

5.2.1 Construction algorithms

Construction algorithms are considered to be the simplest and oldest heuristic approaches to solve the QAP, from a conceptual and an implementation point of view. A construction algorithm consists of successive selection and placement of facilities until a complete layout is achieved. These methods are probably the oldest ones, dating back to the early 60s. The simplicity of construction algorithm is often associated with poor quality of the resulting solutions.

But these construction algorithms can be used to provide initial solutions for improvement algorithms. Improvement methods start with a feasible solution and try to improve it by interchanges of single assignments.

5.2.2 Improvement algorithms

An improvement algorithm starts with an initial solution (existing layout). This existing layout is improved by exchanging the locations of a pair of facilities. The exchange, which produces the best solution, is retained and the procedure continues until the solution cannot be improved any further or until a stopping criterion is reached. Hence, the solution quality of improvement algorithms greatly depends on the initial layout provided, and the systematic procedure of the location exchange.

The greedy nature of pair-wise exchange makes it susceptible to converge to a local optimum. Therefore, the shortcomings of improvement algorithms originate not only from the initial solution provided but also from the greedy nature of the systematic exchange procedure. The greedy nature of the procedure is exposed because only the location exchanges, which result in the greatest cost reduction, are accepted. Hence, the nature of the exchange procedure often impedes the algorithm from finding the global optimum and causes the algorithm to converge to a local optimum.

Improvement methods can easily be combined with construction methods.

Improvement algorithms can be meta-heuristic such as SA and TS, which require one feasible solution as starting solution for the execution of these algorithms.

5.2.3 Hybrid algorithms

In hybrid algorithms the solution of QAP is determined by using a combination of two optimal or sub-optimal algorithms. Such combination of algorithms is essential in some cases to improve solution quality. This classification is extended to include certain algorithms, which use the principal of construction algorithms and improvement algorithms. FLAC and DISCON are examples of such hybrid algorithms.

5.2.4 Graph theoretic algorithms

Graph theoretic algorithms identify maximal planar subgroups of a weighted graph that show the relationships between the facilities [19]. The dual of a maximal planar sub graph determines the layout of the facilities. Seppanen and Moore proposed graph theoretic solutions procedure in which a heuristic algorithm, which uses this strategy, was also presented. The algorithm determines the maximum spanning tree based on the weighted graph. With the help of one edge adding process, the maximum spanning tree is the used to obtain a maximal planar sub graph. The dual of the maximal planar sub graph determines a layout of the facilities.

5.3 Meta-heuristic algorithms

The development of meta-heuristic algorithms has greatly influenced the performance of improvement algorithm and uses a general strategy like pair-wise exchange heuristic. There are three classes widely used of meta-heuristic algorithms in layout problem i.e. Simulated annealing (SA), tabu search (TS), and genetic algorithms (GA).

5.3.1 Simulated annealing algorithms (SA)

Simulated annealing (SA) is a general probabilistic local search algorithm, proposed by Kirkpatrick et al in 1983, to solve difficult optimization problems. Many large instances of difficult real life problems were successfully solved by simulated annealing algorithms. Its ease of implementation, convergence properties and its use of hill-climbing moves to escape local optima has made it a popular technique over two decades. SA is based on the analogy between the annealing of solids and the solving of combinatorial optimization problems [20]. SA is a step-by-step method which could be considered as an improvement of the local optimization algorithm. This process accepts not only better solutions but also worse solutions with a certain probability which is called the probability of accepting. The probability of accepting is determined by the temperature. The probability of accepting a worse solution is large at a higher temperature. As the temperature decreases, the probability of accepting a worse solution also decreases as well.

SA has advantages and disadvantages compared to other global optimization techniques, such as genetic algorithms, tabu search algorithms, and neural networks algorithms. Among its advantages are the relative ease of implementation and the ability to provide reasonably good solutions for many combinatorial problems. Though a robust technique, its drawbacks include the need for a great deal of computer time for many runs and carefully chosen tunable parameters.

5.3.2 Genetic algorithms (GA)

Genetic algorithms (GA) is a heuristic search that mimics the process of natural evolution, which encode a potential solution to a specific problem on a simple chromosome-like data structure and apply operators like mutation, recombination to create new data strings and to preserve critical information [21,22].

GA gained more attention during the last decade than any other evolutionary computation algorithms; it utilizes a binary coding of individuals as fixed-length strings over the alphabet {0, 1}.

Evolution, or more specifically biological evolution, is the change over time in one or more inherited traits of individuals. Natural selection, genetic drift, mutation, gene flow are the four corresponding common mechanisms of evolution. After a long enough time, only the adaptive individuals survive as a consequence of natural selection. To put it concisely, whether the individual should survive or not is decided by two factors, the gene in the individual and the fitness of the gene in the whole population. Mimicking the mechanism, genetic algorithm applies as a searching tool finding out the fittest individuals among a population. More often, the algorithm is viewed as a function optimizer, implementing first by defining two attributes of the individuals: the gene (a data string specifies the individual's character) and the fitness (a function evaluates the individual's vitality). Thus, the two main components of most genetic algorithms that are problem dependent are: the problem encoding (the gene of the individual) and the evaluation function (the fitness) Subsequently, certain operators like mutation and recombination are applied to select the fittest offspring after several generations as the finial individuals, the corresponding optimal answer to the problem when decoded.

GA iteratively search the global optimum, without exhausting the solution space, in a parallel process starting from a small set of feasible solutions and generating the new solutions in some random fashion.

5.3.3 Tabu search algorithms (TS)

Tabu search (TS) was proposed by Glover and has quickly become one of the most effective methods using local search techniques to find near-optimal solution to combinatorial optimization problems. It uses a deterministic local search technique which is able to escape local optima by using a list of prohibited neighbor solutions known as the tabu list. In addition to escaping local optima, using the tabu list can also prevent cycling by forbidding or penalizing moves which take the solution, in the next iteration, to points in the solution space previously visited, and thus save computational time.

A drawback of tabu search is that if it reaches a previously visited solution, it will cycle following the same path unless a tabu neighbor exists. In other words, if the search moves to a previously visited solution that has not been tabu for the last two iterations, then a loop is encountered.

5.3.4 Ant colony algorithms (ACO)

Recently, a few papers have appeared where an ant colony algorithm (ACO) has been attempted to solve large FLP. The first ACO system was introduced by Marco Dorigo in his Ph.D. thesis in 1992, and was called ant system.

ACO is a heuristic search technique to seek for an optimal path in a graph, inspired by the ability of ants to find food sources by using a substance called pheromone. Ant belongs to a colony leave the nest and randomly search for a food source. When an ant finds a food source, it returns to the nest to let others know about the source. On the way back to the nest, the ant places pheromone, which ants are sensitive to, to mark the path from the food source to the nest. Ants select their path to food sources according to the pheromone concentration on different paths. Pheromone evaporates over time, and this causes less frequently visited food sources to lose their address hence to be less visited by others ants. This mechanism has been the cornerstone to devise meta-heuristic algorithms for finding good solutions for difficult optimization problems.

5.4 Other approaches

The major drawbacks of the aforementioned approaches lie in the fact that the search for the best layout is not very efficient and the multi-objective nature are not considered in the problem. As a matter of fact, facility layout problem can be considered one of the truly difficult ill-structured, multi-criteria and combinatorial optimization problems. Many researchers still finding out for new and recent developments rather than conventional approaches to overcome the aforementioned drawbacks. Intelligent techniques such as expert systems, fuzzy logic and neural networks have been used as new advancements for the tackled problem.

6. Computer simulations

The typical absence of some encompassing, closed-form, and analytical fitness functions renders computer simulations a useful alternative. Such an approach would provide detailed analysis, modeling, and evaluation of complex layout design problems. However, simulation models are not easily amenable to optimization and make procurement of a superior layout alternative difficult to achieve. Recently, some efforts have been made to optimize layout design simulation models using genetic algorithms in various facility layout design contexts in order to expedite the process and procure a diverse set of superior in layout alternatives. Nevertheless, computer simulations are usually very time consuming and could become prohibitive in the facility layout design process.

7. Facility layout based on manufacturing costs

Facility layout is composed of product, its process routing, machine and some space. Different combinations of these entities and their activities affect the type of facility layout. Considering the criteria of material handling route, the types of facility layout are classified into three types: single-row layout, multi-row layout and loop layout, as shown in Fig. 6. The application scope, advantages and disadvantages are illustrated in Table 7. The single-row layout includes three shapes such as linear, U-shape and semi-circular. In the linear layout, there may exist bypassing and backtracking, as shown in Fig. 1(a). Backtracking is the movement of some parts from a machine to another machine that precedes it in the sequence of placed machines in a flow line arrangement. Bypassing occurs when a part skips some machines while it is moving towards the end of a flow line arrangement. Table 8 gives the comparison of main characteristics for backtracking and bypassing.

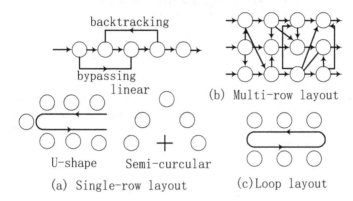

Fig. 6. Types of facility layout based on the criteria of material handling route

Type of layout		Application scope	Advantages	Disadvantages
Single-row layout	Linear	Within GT cells, in facilities that implement JIT, and sometimes with FMS	Material flow are moving along the sequence of operations of all the parts; small material handling cost and time; less delays; better control of operations; the ability to use conveyors.	When several parts having different sequence of operations are processed, the benefits of a flow line arrangement are reduced since the movement of parts may not always be unidirectional.
	U-shape			
	Semi-circular			
Multi-row layout		Suitable for FMS	Adjacent lines share common equipments; low investment; small space area; high machine utilization rate;	Complicated process management; coordinate multi-task difficulty.
Loop layout		Used in FMS	High flexibility in material handling system	

Table 7. Comparison of three types of layout

	Direction	Derivation	Disadvantages	Objective	Scope
Backtracking	Adverse sequence of operations in the flow line	The difference in the sequences of operations of the parts	Impacts the movement cost and productivity of facility	Should be minimize.	In traditional facilities
Bypassing	Same sequence of operations in the flow line	The same with above.	Unnecessary travel time and cost	The same with above	The same with above.

Table 8. Comparison of backtracking and bypassing

7.1 Problem statement

Selling price of products is the concerned problem for customers. Therefore how to decrease the selling price by effective layout planning is an important issue.

7.1.1 Manufacturing cost

Skinner provides the breakdown of costs for a manufacturing product [17], shown in Fig. 7. About 40% of the selling price of a product is manufacturing cost. Material and parts make up the largest percentage of total manufacturing cost, at round 50%. Direct labor is responsible for operating the facilities and is a relatively small proportion of total manufacturing cost: 12%. It is only about 5% of selling price. Machinery, plant and energy etc. are about 26% of manufacturing cost. Therefore, decreasing the manufacturing cost is the key to lower the selling price of products.

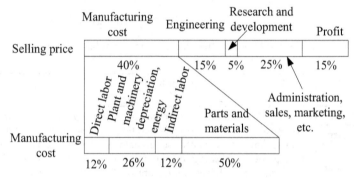

Fig. 7. Breakdown of costs for a manufactured product

Manufacturing cost has two classify methods. The first one classifies the manufacturing cost into fixed costs and variable costs. The second separates manufacturing cost into: (1) direct labor, (2) material, and (3) overhead. In this chapter, the second one is selected. The classification of manufacturing cost is shown in Fig. 8. The direct labor cost is the sum of the wages and benefits paid to the direct labor. The smaller the number of direct labor, the lower the manufacturing cost. The material cost is the cost of all raw material used to manufacture the parts or products. Overhead costs are all of the other expenses associated with running the manufacturing firm. Overhead divides into two categories: (1) factory overhead and (2) corporate overhead. Detail expenses of overhead costs are listed in Table 9.

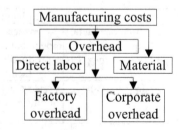

Fig. 8. Classification of manufacturing cost

Factory overhead costs		Corporate overhead costs	
Plant supervision	Applicable taxes	Corporate executives	Applicable taxes
Line foreman	Insurance	Sales and marketing	Cost of space
Maintenance crew	Heat and air conditioning	Accounting department	Security personnel
Custodial services	Light	Finance department	Heat and air conditioning
Security personnel	Power for machinery	Legal counsel	Light
Tool crib attendant	Factory depreciation	Engineering	Insurance
Material handling	Equipment depreciation	Research and development	Fringe benefits
Shipping and receiving	Fringe benefits	Other support personnel	Other office costs

Table 9. Typical overhead costs

Factory overhead consists of the costs of operating the factory other than direct labor and materials. Corporate overhead is the cost of running the company other than its manufacturing activities. As shown in Table 9, material handling cost in factory overhead and cost of space in corporate overhead are the two parts which relate to the facility layout. When material handling cost and cost of office space increase, the manufacturing cost increase accordingly.

7.1.2 Objectives of facility layout based on manufacturing cost

Groover observes that materials spend more time waiting or handling than in process [23]. His observation is illustrated in Fig. 9. Only 5% of the time is spent on the machine. About 95% of a part's time is spent either moving or waiting. This figure shows that the material handling and storage are significance in a typical factory. Furthermore, studies show that material handling cost makes up 20%-50% of the total operating cost and 15%-70% of the total manufacturing cost. Therefore, the most common objective of facility layout is the minimization of material handling cost.

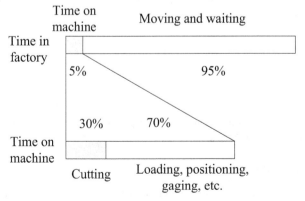

Fig. 9. How time is spent by a typical part in batch production machine shop

When the cost of space increases, the manufacturing cost increase accordingly. With the given purchase price of unit space area, the higher the area utilization rate is, the lower the manufacturing cost is. So, one objective of facility layout is to maximize the area utilization rate.

Combination of the Fig. 7 and 8 show that other than materials and parts, direct labor and machinery are main parts of manufacturing cost. The higher the utilization rate of direct labor and machinery are, the lower the manufacturing cost is. Increased utilization of existing machinery could lead to smaller machine inventories since less machinery would be sitting unused, and the direct labor are the same. Hence the maximization of the utilization rate of direct labor and machinery are also the objective of facility layout.

To sum up, the objectives of facility layout based on manufacturing cost include: (1) minimizing the material handling cost, (2) maximizing the area utilization rate, and (3) maximizing the utilization rate of direct labor and machinery.

7.2 Model formulation

7.2.1 Material handling cost

For manufacturing facilities, material handling cost is the most significant measure for determining the efficiency of a layout and is most often considered. It is determined based on the flows of materials between departments and the distances between the locations of the departments. The material handling cost model has the following form:

$$F = \sum_{i=1}^{n}\sum_{j=1}^{n} c_{ij} f_{ij} d_{ij} \qquad i,j = 1,2,...,n \qquad (20)$$

Note that c_{ij} is the unit cost (the cost to move one unit load one distance from department i to j), f_{ij} is the material flow between the department i and j, d_{ij} is the distance between the centers of department i and j.

7.2.2 Area utilization rate

Area utilization rate of whole layout is a ratio of total areas required of all facilities to the smallest possible rectangle, which can envelop all the facilities [24]. Hence, the area utilization area rate of whole layout is shown as follow:

$$R_s = \frac{\sum_{i=1}^{n} A_i}{\sum_{i=1}^{n} A_i + \sum B_j} \times 100\% \qquad (21)$$

Note that R_s is the Area utilization rate, A_i is the area of department i where equipment i is sitting, B_j is the blank area of layout.

7.2.3 Equipment utilization

Up to now, there exist three views about equipment utilization. Østbye defines that equipment utilization is measured as of the ratio of the number of units reported in use by the surveyor relative to the total number of units recorded as present by the surveyor [25]. This measure was calculated daily, for both the intervention and control wards, during the pre- and post-intervention control periods as well as the intervention period. Optimal efficiency of utilization would have a ratio of 1, indicating that every unit present was in use. Increased utilization of existing equipment could lead to smaller equipment inventories since less equipment would be sitting unused. Michael Vineyard thinks that equipment utilization measures the percentage of time the machines are in use and considers factors beyond just maintenance downtime [26]. Steege puts forward that equipment utilization has three affecting factor: rate of quality, availability and performance efficiency [27]. Rate of quality measures the percentage of defect-free product that is manufactured by a piece of equipment. It determines the effect of the equipment on product yield. It is equally significant from a productivity point of view whether or not the equipment is running at full capacity. Equipment availability measures the percentage of time that equipment is ready to perform its manufacturing function. Performance efficiency is the percentage of available time that equipment is producing sellable product.

In this chapter, performance efficiency is selected to evaluate the equipment utilization. The equipment utilization affected by performance efficiency can be formally stated as follow:

$$R_{EU} = \frac{T_O}{T_A} \times 100\% \tag{22}$$

Note that R_{EU} is equipment utilization, T_O is the operation time of equipment, including processing time, unload time, and setup time, T_A is the available time of equipment.

7.2.4 Labor utilization

Labor utilization measures as average hours worked through overtime work (and possibly short-time work) [28]. The optimal labor utilization, i.e. hours worked per employee is explained by average wage rates which are functions of the hours worked [29]. Thus, Average labor utilization of layout can be written as:

$$R_L = \frac{T_W}{T_T} \times 100\% \tag{23}$$

Note that R_L is average labor utilization, T_W is the work time of labor, including processing, loading, unloading, loaded and empty travel time, T_T is the total time of labor in a factory.

7.3 Simulation and results

The simulation was carried out in Deneb/QUEST platform in order to investigate the performance of three types of facility layout. QUEST is a discrete event simulation software package, used to model and simulate the operation of complex automated manufacturing systems. Using 3D CAD geometry, QUEST analyzes the performance of existing or proposed manufacturing facilities by simulating the process behavior over a specified time. QUEST combines a graphical user interface with material flow logic grouped in modules for: labor, conveyors, automated guided vehicles (AGVs), kinematics, power and free conveyors, and automated storage and retrieval systems (AS/RS). A Value-Added Costing module assists in implementation of Activity Based Costing during the simulation analysis, Statistical results can be viewed with graphical and numerical analysis capabilities.

The piston production line-117 is chosen as an example to simulate. The simulation procedure is illustrated in Fig. 10. Virtual facility model can be gain from the equipment database of QUEST. The process parameters of piston are inputted to QUEST through process DB. Logic and algorithm DB provide rules and procedures that govern the behavior of the element and algorithm for the simulation system. At the final of this procedure, QUEST supplies simulation data to user in order to analysis the three types of facility layout.

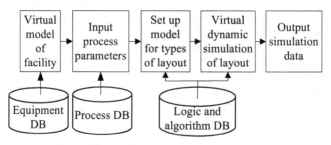

Fig. 10. Simulation procedure of facility layout

7.3.1 Assumptions and constrains

Two major assumptions made in the proposed models are as follows: (1) Machines are rectangular with the same dimensions and the distance between the machines is calculated with respect to their centers. (2) The clearance between each pair of machines is fixed.

For three types of facility layout, two set of constrains are considered: (1) one machine is assigned to each location and each machine is assign to only one location; (2) the clearance between each pair of macines has the minimal value to avoid intervening and overlapping with each other.

Based on assumptions and constrains, the simulation models of the facility layout described above are given in Fig. 11.

7.3.2 Simulation results and analysis

The man-hour arrangement of piston production line is listed in Table 10. Quantity represents the number of machines. Setting the simulation time is one work day, i.e. 6.5 hours. The loaded travel time is measured by minute. Setting the moving velocity of labor is 304mm/sec and c_{ij} is 1. The simulation results are compared in Table 11~14.

The simulation results listed in Table 11 show that the material handling cost of loop layout is the lowest, and multi-line layout is the highest due to the distances between the centers of departments in which the materials are handled. As for area utilization rate shown in Table 12, loop layout is higher than the others, and semi-circular layout is the lowest of the three types of facility layout as a result of the blank areas of it is the largest. The results in Fig. 7 show that the equipment utilization of U-shape layout is the higher than others in the same moment. At aspect of labor utilization presented in Fig. 8, the U-shape layout is better than the others, and linear layout is the worst owing to the sum of its hanling materials is minimum.

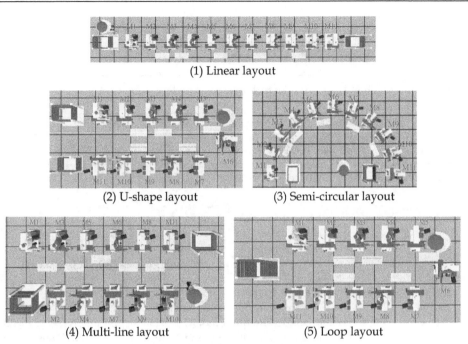

(1) Linear layout

(2) U-shape layout

(3) Semi-circular layout

(4) Multi-line layout

(5) Loop layout

Fig. 11. Simulation model of three types of facility layout

No	Machine	Quantity	Labor	Man-hour(s)	No	Machine	Quantity	Labor	Man-hour(s)
1	M1	1	L1	43.28 (dual-workstation)	6	M6	1	L4	31.3
2	M2	1	L1	74.3	7	M7	1	L4	50.26
3	M3	1	L2	74.3	8.9	M8,9	2	L5	136.2
4	M4	1	L3	36	10	M10	1	L6	70.77
5	M5	1	L3	36	11	M11	1	L6	14.7

Table 10. Man-hour arrangement of piston production line

Type of facility layout	Linear layout	U-shape layout	Semi-circular layout	Multi-line layout	Loop layout
Material handling cost	688.607	723.2155	681.5551	880.5325	662.1082

Table 11. Material handling cost of piston production line (/yuan)

Type of facility layout	Linear layout	U-shape layout	Semi-circular layout	Multi-line layout	Loop layout
Material handling cost	30	48.26	19.096	48.26	66.47

Table 12. Area utilization rate of piston production line (%)

Type of facility layout	Linear layout	U-shape layout	Semi-circular layout	Multi-line layout	Loop layout
Equipment utilization rate	47.19764	51.43864	47.22245	46.719	45.50791

Table 13. Equipment utilization rate of piston production line (%)

Type of facility layout	Linear layout	U-shape layout	Semi-circular layout	Multi-line layout	Loop layout
Labor utilization rate	24.30267	32.38983	25.89	29.1795	27.006

Table 14. Labor utilization rate of piston production line (%)

7.4 Conclusions

In this example, single-row layout, multi-row layout and loop layout are compared with each other first. The backtracking and bypassing in linear layout are introduced. Due to customers' demand for as possible as low selling price of a product, the concept of manufacturing cost is presented secondly. Based on analyzing the components of manufacturing cost, four models are established lately. These models describe the function of material handling cost, area utilization rate, equipment utilization, and labor utilization, respectively. Through modeling and simulation on QUEST platform, three types of facility layout (linear, U-shape, semi-circular, multi-line, and loop) are compared finally. From above discussion, some conclusions can be achieved as follows: (1) material handling cost of loop layout is lowest among the three types of facility layout; (2) area utilization rate of loop layout is higher than the others; (3) equipment utilization of U-shape layout is the higher than others in the same moment; (4) labor utilization of U-shape layout is highest, and linear layout is the worst.

8. References

[1] A.R. McKendall, J. Shang. Hybrid ant systems for the dynamic facility layout problem. Computers & Operations Research, 2006, 33(3) , pp.790
[2] P. Kouvelis, A.A. Kurawarwala and G. Gutierrez. Algorithms for robust single and multiple period layout planning for manufacturing systems. Eur. J. Oper. Res., 1992, 63(2) , pp.287
[3] A.L. Page. PDMA new product development survey: Performance and best practices. Paper presented at PDMA Conference, 1991, 13
[4] B. Jaydeep, H.C. Chun and G.C. Daniel. An improved pair-wise exchange heuristic for the dynamic plant layout problem. INT. J. PROD. RES.,2000, 38(13) , pp.3067
[5] T. Yang, B.A. Peters. Flexible machine layout design for dynamic and uncertain production environments. Eur.J.Oper.Res., 1998,108(1) , pp.49
[6] T. Ertay, D. Ruan and U.R. Tuzkaya. Integrating data envelopment analysis and analytic hierarchy for the facility layout design in manufacturing systems. Information Sciences, 2006,176(3) , pp.237
[7] M.J. Rosenblatt, H.L. Lee. A robustness approach to facilities design. INT. J. PROD.RES., 1987, 25(4) , pp.479

[8] T.L. Urban. Computational performance and efficiency of lower-bound procedures for the dynamic facility layout problem. Eur.J.Oper.Res., 1992,57(2) , pp.271

[9] J.A. Tompkins, J. A. White and Y. A. Bozer. Facilities planning. NewYork: Wiley; 1996, pp.137

[10] R.D. Meller, K.Y. Gau. The facility layout problem: recent and emerging trends and perspectives. J. Manuf. Systems, 1996, 15(5),pp.351

[11] R.L. Francis, J.A. White. Facility layout and location. New Jersey: Prentice Hall, Englewood Cliffs, 1974

[12] Koopmans T. C., Beckman M. Assignment problems and the location of economic activities. Econometrica, 1957, 25(5),pp.53

[13] Bazaraa M. S.. Computerized layout design: a branch and bound approach. AIIE Transactions, 1975, 7(4) ,pp.432

[14] Lawler E. L.. The quadratic assignment problem[J]. Management Science, 1963, 9(4), pp.586

[15] Kaufman L., Broeckx F.. An algorithm for the quadratic assignment problem using Benders'decomposition. European Journal of Operational Research, 1978, 2(3) ,pp.207

[16] Foulds L. R., Robinson D. F.. Graph theoretic heuristics for the plant layout problem. Operations Research Quarterly, 1976, 27(4) ,pp.845

[17] W. Skinner. The focused factory. Harvard Business Review, May-June 1974,pp.113

[18] Heragu S.S.. Recent models and techniques for solving the layout problem. European Journal of Operational Research, 1992, 57(2) , pp.136

[19] Seppannen J.J., Moore J.M.. Facilities Planning with Graph Theory. Management Science, 1970, 17(4):242-253.49. Seppannen J.J., Moore J.M.. String Processing Algorithms for Plant Layout Problems. Int. J. Prod. Res., 1975, 13(3) , pp.239

[20] Elmohamed S., Fox G., and Coddington P.. A comparison of annealing techniques for academic course scheduling. Proceedings of the 2nd international conference on the practice and theory of automated timetabling, 1998, New York, pp.146

[21] Azadivar F., G. Tompkins. Simulation Optimization with Quantitative Variables and Structural Model Changes: A Genetic Algorithm Approach. Euro. J. of Op. Res., 1999, 113, pp.169

[22] Azadivar F., J. Wang. Facility layout Optimization using Simulation and Genetic Algorithms. International Journal of Production Research, 2000, 38(17) , pp.4369

[23] M.P. Groover. Fundamentals of Modem Manufacturing: Materials, Processes, and Systems. New York, John Wiley & Sons, Inc.,1996

[24] M.J. Wang. A solution to the unequal area facilities layout problem by genetic algorithm. Computers in Industry, Vol.56(2005),pp.207

[25] T. Østbye, D.F. Lobach, D. Cheesborough, A.M.M. Lee, K.M. Krause, V. Hasselblad, and D. Bright, "Evaluation of an infrared/radiofrequency equipment-tracking system in a tertiary care hospital" , Journal of Medical Systems, 2003,27(4),pp.367

[26] M. Vineyard, A.G. Kwasi, J.R. Meredith, "An evaluation of maintenance policies for flexible manufacturing systems: A case study", Int. J. Oper. & Prod. Mgmt., 2000,20(4),pp.409

[27] P. Steege, "Overall equipment effectiveness in resist processing equipment", Advanced Semiconductor Manufacturing Conference. IEEEISEMI, 1996

[28] H. Entorf, H. König, W. Pohlmeier, "Labor utilization and nonwage labor costs in a disequilibrium macro framework", Scand. J. of Economics, 1992, 94(1),pp.71

[29] T. Airaksinen. On Optimal Utilization of Labor and Capital Stocks in the Neoclassical Theory of the Firm. Scand. J. of Economics, 1977, pp.28

10

Real-Time Petri Net Based Control System Design for Distributed Autonomous Robotic Manufacturing Systems

Gen'ichi Yasuda

Nagasaki Institute of Applied Science,

Japan

1. Introduction

Generally speaking, flexible manufacturing systems are made up of some flexible production machines with some local storage facilities for tools and parts, some handling devices such as robots and a versatile transportation system. It is expected that more and more robots will be introduced into manufacturing systems to automate various operations in the near future. However, it is quite obvious that a single robot cannot perform effective tasks in an industrial environment, unless it is provided with some additional equipment that allows the machine to grasp, handle and dispose correctly workpieces or mechanical parts onto which technological operations are to be performed. Therefore, in order to avoid the need of loading and unloading of parts to the robot manually, it is usually required to integrate the robot into the production line that also includes machine tools, conveyors, and other special purpose machines. Mainly to provide flexibility to robots, a lot of researches have been done to develop an effective programming method for robots. But not much research has been done to integrate a system which includes various machines (robots and other devices) that cooperate in the same task (Holding & Sagoo, 1992). A common programming language for tasks that involve more than one robot or machine should be provided (Holt & Rodd, 1994).

Robot programs often must interact with people or machines, such as feeders, belt conveyors, machine tools, and other robots. These external processes are executing in parallel and asynchronously; therefore, it is not possible to predict exactly when events of interest to the robot program may occur. The programmable logic controllers (PLC) are widely used to the programming and control of flexible manufacturing systems. Implementation languages can be based on ladder diagrams or more recently state machines. However, when the local control is of greater complexity, the above kinds of languages may not be well adapted. It is important to have a formal tool powerful enough to develop validation procedures before implementation. Conventional specification languages such as ladder diagrams do not allow an analytical validation. Presently, the implementation of such control systems makes a large use of microcomputers. Real-time executives are available with complete sets of synchronization and communication primitives (Yasuda, 2000). However, coding the specifications is a hazardous work and debugging the implementation is particularly difficult when the concurrency is important.

The Petri net and its graphical representation is one of the effective means to describe control specifications for manufacturing systems. From the plant control perspective, the role and the presence of nets were considered in the scheduling, the coordination and the local control level (Silva, 1990). However, in the field of flexible manufacturing cells, the network model becomes complicated and it lacks the readability and comprehensibility. Therefore, the flexibility and expandability are not satisfactory in order to deal with the specification change of the control system. Despite the advantages offered by Petri nets, the synthesis, correction, updating, etc. of the system model and programming of the controllers are not simple tasks. The merging of Petri nets and knowledge based techniques seems to be very promising to deal with large complex discrete event dynamic systems such as flexible manufacturing systems (Gentina & Corbeel, 1987; Maletz, 1983; Wang & Sarides, 1990).

The aim of this chapter is to introduce manufacturing engineering specialists to the basic system level issues brought up by the development of computer-controlled robotic manufacturing systems and how Petri nets are applied to resolve the above mentioned problems of control system design. After some terminology concerning basic Petri nets, the extensions of Petri nets for manufacturing system control are briefly reviewed. Based on the hierarchical and distributed structure of the manufacturing system, the net model of the system is decomposed into a set of interacting local nets and a system coordinator net to perform distributed autonomous multitasking control based on Petri nets.

2. Modeling of discrete event manufacturing systems with Petri nets

The Petri net is one of the effective means to represent discrete event manufacturing systems. Considering not only the modeling of the systems but also the well-defined control, the guarantee of safeness and the capabilities to represent input and output functions are required. Therefore the Petri net has been modified and extended.

2.1 Modification of basic Petri nets

A Petri net is a directed graph whose nodes are places shown by circles and transitions shown by bars. Directed arcs connect places to transitions and transitions to places. Formally, a Petri net is a bipartite graph represented by the 4-tuple $G = \{P,T,I,O\}$ (Murata, 1989) such that:

$P = \{p_1, p_2, ..., p_n\}$ is a finite, not empty, set of places;

$T = \{t_1, t_2, ..., t_m\}$ is a finite, not empty, set of transitions;

$P \cap T = \phi$, i.e. the sets P and T are disjointed;

$I : T \rightarrow P^\infty$ is the input function, a mapping from transitions to bags of places;

$O : T \rightarrow P^\infty$ is the output function, a mapping from transitions to bags of places.

The input function I maps from a transition t_j to a collection of places $I(t_j)$, known as input places of a transition. The output function O maps from a transition t_j to a collection of places $O(t_j)$, known as output places of a transition.

Each place contains integer (positive or zero) marks or tokens. The number of tokens in each place is defined by the marked vector or marking $M = (m_1, m_2, ..., m_n)^T$. The number of tokens in one place p_i is simply indicated by $M(p_i)$. The marking is shown by dots in the places. The marking at a certain moment defines the state of the net, or the state of the system described by the net. The evolution of the state therefore corresponds to an evolution of the marking, caused by the firing of transitions. The firing of an enabled transition will change the token distribution (marking) in a net according to the transition rule. In a basic Petri net, a transition t_j is enabled if $\forall p_i \in I(t_j)$, $M_k(p_i) \geq w(p_i, t_j)$, where the current marking is M_k and $w(p_i, t_j)$ is the weight of the arc from p_i to t_j.

Because discrete event manufacturing systems are characterized by the occurrence of events and changing conditions, the Petri net type considered is the condition-event net, in which conditions can be modeled by places whilst events can be modeled by transitions. Events are actions occurring in a system. The occurrence of these events is controlled by system states. Because the condition-event system is essentially asynchronous, events always occur when their conditions are satisfied. Consequently, bumping occurs when despite the holding of a condition, the preceding event occurs. This can result in the multiple holding of that condition. From the viewpoint of discrete event process control, bumping phenomena should be excluded. So, the firing rule of the basic Petri net should be modified so that the system is free of this phenomenon. Thus the axioms of the modified Petri net are as follows:

1. A transition t_j is enabled if for each place $p_k \in I(t_j)$, $m_k = 1$ and for each place $p_l \in O(t_j)$, $m_l = 0$;
2. When an enabled transition t_j is fired, the marking M is changed to M', where for each place $p_k \in I(t_j)$, $m'_k = 0$ and for each place $p_l \in O(t_j)$, $m'_l = 1$;
3. In any initial marking, there must not exist more than one token in each place.

The number of arcs terminated at or started from a place or a transition is unlimited, but at most one arc is allowed between a transition and a place. According to these axioms, the number of tokens in each place never exceeds one, thus, the modified Petri net is said to be a safe graph. The modified Petri net is a subclass of the Petri net, and it is transformed into the equivalent Petri net as shown in Fig. 1.

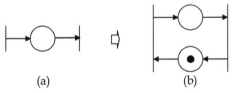

(a)   (b)

Fig. 1. (a) A place in the modified Petri net and (b) its equivalent Petri net

2.2 Extensions for real-time control

The extended Petri net adopts the following elements as input and output interfaces which connect the net to its environment: gate arc and output signal arc. A gate arc connects a transition with a signal source, and depending on the signal, it either permits or inhibits the occurrence of the event which corresponds to the connected transition. Gate arcs are classified as permissive or inhibitive, and internal or external. When the signal is 1 (true), a permissive arc permits the occurrence of the event. On the other hand, an inhibitive arc inhibits the occurrence of the event when the signal is 1. An internal arc deduces the signal from a place, and the signal is 1 when a token exists in the place, otherwise 0 (false). An external arc deduces the signal from an external machine. An output signal arc sends the signal from a place to an external machine. In addition to the axiom 1, a transition is enabled if it does not have any internal permissive arc signaling 0 nor any internal inhibitive arc signaling 1. An enabled transition is fired if it does not have any external permissive arc signaling 0 nor any external inhibitive arc signaling 1. Thus the enabling condition and the external gate condition are formally expressed as follows.

$$t_j = \bigcap_{m=1}^{M} p_{j,m}^I \wedge \bigcap_{n=1}^{N} \overline{p_{j,n}^O} \wedge \bigcap_{q=1}^{Q} g_{j,q}^{IP} \wedge \bigcap_{r=1}^{R} \overline{g_{j,r}^{II}} \tag{1}$$

$$g_j^E = \bigcap_{u=1}^{U} g_{j,u}^{EP} \wedge \bigcap_{v=1}^{V} \overline{g_{j,v}^{EI}} \tag{2}$$

where

M : set of input places of transition j

$p_{j,m}^I$: state of input place m of transition j

N : set of output places of transition j

$p_{j,n}^O$: state of output place n of transition j

Q : set of internal permissive gate signals of transition j

$g_{j,q}^{IP}$: internal permissive gate signal variable q of transition j

R : set of internal inhibitive gate signals of transition j

$g_{j,r}^{II}$: internal inhibitive gate signal variable r of transition j

U : set of external permissive gate signals of transition j

$g_{j,u}^{EP}$: external permissive gate signal variable u of transition j

V : set of external inhibitive gate signals of transition j

$g_{j,v}^{EI}$: external inhibitive gate signal variable v of transition j

All the variables are logical binary variables, and $\wedge$, $\vee$ denote the logical product and the logical sum, respectively, and $\bigcap_{i=1}^{m} a_i = a_1 \wedge a_2 \wedge ... \wedge a_m$. The state (marking) change, that is, the addition or removal of a token of a place, is described as follows:

$$p_{j,m}^I = p_{j,m}^I \wedge \overline{(t_j \wedge g_j^E)} = RST(t_j \wedge g_j^E) \tag{3}$$

$$p_{j,n}^O = p_{j,n}^O \vee (t_j \wedge g_j^E) = SET(t_j \wedge g_j^E) \tag{4}$$

where $SET()$ and $RST()$ denote the set and the reset function, respectively.

Fig. 2 shows an example of extended Petri net model of robotic task control by transition firing with permissive and inhibitive gate arcs. The robot starts the loading operation based on signals from the switches, sends the commands through output signal arcs, and receive the status signals from the sensors through permissive gate arcs. Fig. 3 shows an example detailed net model of the lowest level local control of a machining center.

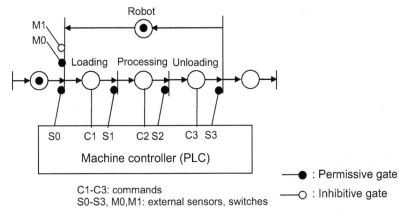

Fig. 2. Extended Petri net representation of robotic task with output signal arcs and gate arcs.

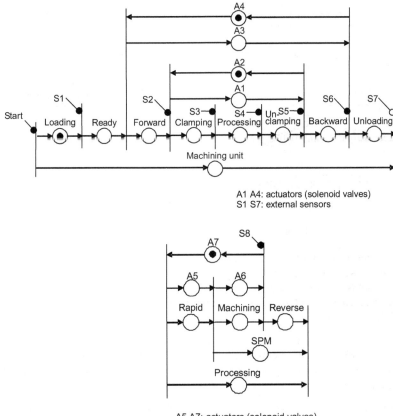

Fig. 3. Detailed net model of real-time control of manufacturing tool

3. Edition and simulation of net models

When programming a specific task, the task is broken down into subtasks. These subtasks are represented by a place. The internal states of machines are also represented by a place. The relations between these places are explicitly represented by interconnections of the transitions, arcs and gates. The whole task is edited with a net edition and simulation system. In parallel a graphic robot motion simulator system is used to edit a subtask program for a robot. The basic edition and simulation procedure is shown in Fig. 4.

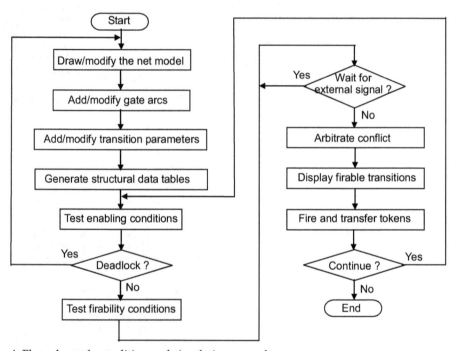

Fig. 4. Flow chart of net edition and simulation procedure

The net simulator is a tool for the study of condition-event systems and used to model condition-event systems through its graphical representation. When the net modeling is finished, the net is transformed into a tabular form and several data tables corresponding to the connection structure of the net are automatically generated (Yasuda, 2008). These tables are the following ones:

1. The table of the labels of the input and output places for each transition;
2. The table of the transitions which are likely to be arbitrated for each conflict place;
3. The table of the gate arcs which are internal or external, permissive or inhibitive, for each transition.

Although a variety of software implementations of Petri nets is possible using multitask processing (Taubner, 1988), a simple implementation method is adopted, where just one process is provided for the management of all places and tokens. Through the simulation steps, the transition vector table is efficiently used to extract enabled or fired transitions.

The table of marking indicates the current marking for each place. Using these data tables, the flow of the net simulation consists in the following steps:

1. Search enabled transitions using the axiom 1 or (1);
2. Test the enabled transitions considering gate conditions (2);
3. Arbitrate enabled transitions in conflict using some arbitration rule;
4. Execute transition firing and output corresponding signals to external machines;
5. Change the marking to the new marking using the axiom 2 or (3), (4) and update the system state.

The flow chart of the enabling condition test is shown in Fig. 5. The simulation algorithm is based on the execution rules of the net. The simulator tests each transition as to whether its input and output places and its internal gate arcs satisfy the enabling condition. If there is no enabled transition, it means that the net is in a deadlock condition. The simulator warns and requires the operator to change the initial marking or structure of the net. If there are some enabled transitions, it tests each of them as to whether its external gate arcs satisfy the firability condition, as shown in Fig. 6. If there is no firable transition, the simulator stops and shows which transitions are waiting for the gate signals.

For an example net as shown in Fig, 7, the enabling condition and the firability condition are written as (5), (6), respectively. The simulator tests each transition in the specified order of (5), (6). Fired transitions are memorized, and through their output places the output transitions of each place are searched. The enabling condition test is performed only for these transitions in order to shorten computation time. In Fig. 7, the enabling condition of only the transition t1 is evaluated, since the transition t5 is fired previously.

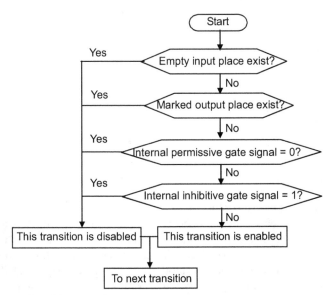

Fig. 5. Flow chart of enabling condition test

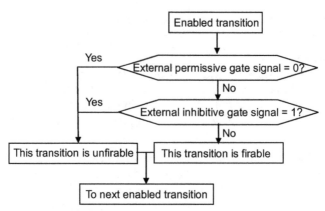

Fig. 6. Flow chart of firability condition test

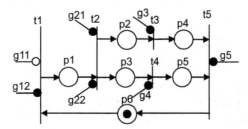

Fig. 7. Example of net representation with parallel activities.

$$t_1 = p_6 \cdot \overline{p_1}$$
$$t_2 = p_1 \cdot \overline{p_2} \cdot \overline{p_3}$$
$$t_3 = p_2 \cdot \overline{p_4}$$

$$t_4 = p_3 \cdot \overline{p_5}$$
$$t_5 = p_4 \cdot p_5 \cdot \overline{p_6}$$

$$(5)$$

$$p_6 = RST(t_1 \cdot \overline{g_{11}} \cdot g_{12})$$
$$p_1 = SET(t_1 \cdot \overline{g_{11}} \cdot g_{12})$$
$$p_1 = RST(t_2 \cdot g_{21} \cdot g_{22})$$
$$p_2 = SET(t_2 \cdot g_{21} \cdot g_{22})$$
$$p_3 = SET(t_2 \cdot g_{21} \cdot g_{22})$$
$$p_2 = RST(t_3 \cdot g_3)$$

$$p_4 = SET(t_3 \cdot g_3)$$
$$p_3 = RST(t_4 \cdot g_4)$$
$$p_5 = SET(t_4 \cdot g_4)$$
$$p_4 = RST(t_5 \cdot g_5)$$
$$p_5 = RST(t_5 \cdot g_5)$$
$$p_6 = SET(t_5 \cdot g_5)$$

$$(6)$$

If the transitions connected to a conflict place may happen to be in conflict, according to the rules of the net, only one of them is chosen to fire arbitrarily and the others become unfirable. The arbiter assigns the right of the order of firing among the transitions connected to a conflict place. But the right vanishes when the specified transition is not firable. The arbiter has a pointer to memorize the transition to be assigned the right next. The procedure of the arbitration is shown in Fig. 8. After the arbitration, all the firable transitions are displayed and fired. The simulator moves the tokens; it remove tokens in all the input places of the fired transitions and put a token in each output place of the transitions. If some error is found or the simulation result does not satisfy the specification, it can be easily amended by reediting the net and by simulating it again. The edition and simulation are performed in an interactive form on a graphic display. The software written in Visual C# under OS Windows XP allows net models be modified on-line and simulation immediately restarted.

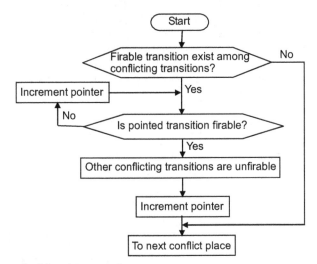

Fig. 8. Flow chart of arbitration procedure

In the basic Petri net, the firing of a transition is indivisible; the firing of a transition has duration of zero. The real-time performance of systems can be studied by adding time to the basic Petri net. An approach known as the timed Petri net associates a time parameter T with a transition, such that once the transition is enabled, it will fire after the period T. If the enabling condition is not satisfied before the schedule time comes, then the transition can not be fired and the passage of time is cancelled. Time values may be associated with places in order to maintain the instantaneous firing rule for transitions. A place with capacitance C_N, such as buffers in manufacturing systems, can be represented as a cascade connection of ordinary places with capacitance 1. The internal gate signal from the place is 1 when the number of tokens in the place is C_N, and 0 when the number is 0. These extensions are illustrated in Fig. 9(a) and (b).

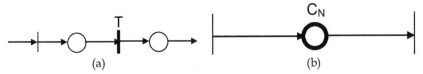

Fig. 9. Example of representation of (a) timed transition (b) place with capacitance N

4. Net models of multitasking control

Manufacturing tasks are a combination of several processes. These processes represent subtasks that are composed of task units. Tasks that include cooperative subtasks of different machines are typical examples of concurrent processes. A system with one process is the degenerate case of a system of concurrent processes, which is obtained by combining nets representing several processes. Every sequential program can be represented by a flow chart. A flow chart is composed of nodes and arcs between them. It represents the flow of control in a program and can be represented by a Petri net, by replacing the nodes with places and the arcs with transitions as shown in Fig. 10. Each arc of the flow chart is represented by exactly one transition in the corresponding net. Petri net models of sequential constructs are shown in Fig. 11. A token residing in a place means that the program counter is positioned ready to execute the next instruction. Places for motion and computational actions have a unique output transition. Decision actions introduce conflict into the net. The choice can either be made nondeterministically or may be controlled by some external signal.

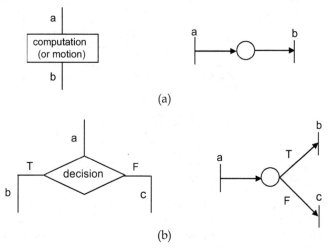

Fig. 10. Translation from nodes in a flow chart to places in a Petri net: (a) computation or motion, (b) decision

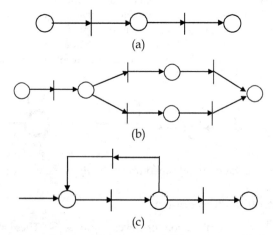

Fig. 11. Net representations of sequential constructs; (a) sequence, (b) decision, (c) iteration

In the case of two concurrent processes, where each process can be represented by a net model of a sequential process, the composite net which is simply the union of such nets can represent the concurrent execution of two processes. Parallelism is usefully introduced into a system only if the component processes can cooperate in the system. Such cooperation requires the sharing of information and resources between the processes. This sharing must be controlled to ensure correct operation of the overall system. One of the most popular synchronization mechanisms has been the P and V operations on semaphores. The WAIT and SIGNAL statements are used in a program written in a high level robot language and provides a variation of the P and V operations as a basic inter-process communication mechanism. Fig. 12 shows the net representation of an example of synchronization mechanism.

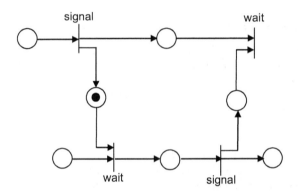

Fig. 12. Net representation of synchronization mechanism using asynchronous communication

Fig. 13 shows the net representation of cooperative operation using synchronization mechanism, where shared transitions require mutual synchronization between two robots. In contrast to decentralized implementation, synchronization can be also implemented by centralized coordination (Yasuda, 2010).

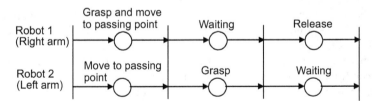

Fig. 13. Net representation of operation that passes a part from right arm to left arm

The main flow of execution control of robotic action using output signal arc and permissive gate arc is described as the following steps:

1. When a token is placed in a place which represents an action, the net based controller initiates the execution of the action (subtask) attached to the fired transition by sending the "start" signal through the output signal arc to the machine controller.

2. Then the machine controller interprets the request and runs the execution routine by sending the commands through serial interface to the robot or other external machine.
3. When the action is completed, the machine controller informs the system controller to proceed with the next activations by sending the "end" signal through external permissive gate arc.

When a token is placed in a place which is "ready" state in the net model, the controller sends the "ready" signal. If the machine receives the signal, it runs the processing routine which performs the initializations and other preliminary processing for the next execution routines. When the processing routine is completed, it sends the "ack" (acknowledgement) signal to the system controller. The "end" and "ack" signals work as gate signals for the system controller.

5. Implementation of real-time control system for robotic cells

To implement the Petri net based modeling and control method, the net based task editor and simulator, and the real-time controller based on tasks represented as net models were developed (Yasuda, 2008). The subtasks and sets of point data needed to execute the whole task are initially identified. Then they are edited and tested with the net based edition and simulation system. Initially, the proposed method is used to execute a simple example of pick-and-place task by a single robot. The experimental set up includes the following equipment: a small industrial robot with an arm (Mitsubishi Electric, Movemaster II RM501), two belt conveyors with their sequence control circuits, a NC machine tool and a general PC. All the software is written in Microsoft Visual C# on Windows XP. The task specification is represented as the flow of a workpiece and written as the following steps:

1. A workpiece arrives at point E1.
2. Conveyor CV1 carries the workpiece to point E2.
3. Robot R1 transfers the workpiece to point E3.
4. Machining operation M1 is done.
5. Robot R1 transfers the workpiece to point E4.
6. Conveyor CV2 carries the workpiece to point E5.

Synchronous cooperation is required to perform the loading and unloading operations between the robot and the conveyor or machining center. The cooperation can be implemented by a system coordinator which coordinates the machine controllers such that associated transitions of the local net models fire simultaneously. For high efficiency, it is desirable that the system accepts as many workpieces as possible, but it must not be in a deadlock condition. Generally, if there are some paths between two transitions, the largest number of tokens in each path is the smallest number of places of the paths. The task specification is shown as follows. Using the place of capacity control, the net representation of the task program written under these requirements is shown in Fig. 14.

Another example is a cooperative task by two arm robots which must synchronize their actions with each other. The task specification is summarized as the following steps:

1. A workpiece arrives at point E1.
2. Robot R1 transfers the workpiece to the exchange area, and at the same time Robot R2 moves to the exchange area.

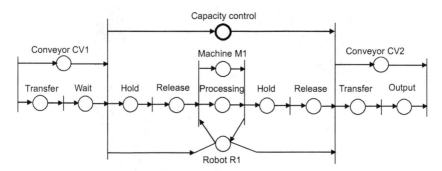

Fig. 14. Net representation of pick-and-place operation with a single robot

3. The workpiece is exchanged from robot R1 to robot R2.
4. Robot R2 changes the workpiece orientation.
5. Robot R2 transfers the workpiece to the exchange area. Robot R1 moves to the exchange area.
6. The workpiece is exchanged from robot R2 to robot R1.
7. Robot R1 transfers the workpiece to point E2.

Following the same procedure of the former example, the subtasks and sets of point data needed to execute the whole task are initially identified. Then they are edited and tested with the net based edition and simulation system. The net representation is written using shared transitions for system coordination as shown in Fig. 15. An experimental view of the cooperative task, passing and exchanging a workpiece, by two robots is shown in Fig. 16.

The detailed procedure of the implemented real-time control based on tasks represented as net models is described as follows. If there is a token in a place corresponding to subtasks, the net based controller sends a message to the respective hardware controllers such as arm, hand, sensor, etc. to execute the defined subtask with certain point data. These parameters (hardware controller code, subtask file code, point data file code) are defined during the net edition procedure. The net based controller was developed with all functions of the edition and simulation to permit correction or modification of the net model on-line. This characteristic is important to facilitate the debugging work. By executing the net model, the

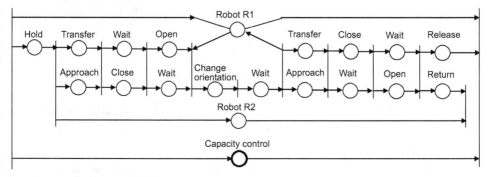

Fig. 15. Detailed net model of cooperative task by two arm robots

Fig. 16. Experiments of cooperative task by two arm robots

developed control system activates the arm, hand, and sensor, etc. and coordinates each individual controller. In making these experiments, it was verified that the implemented system can be used as an effective tool for introducing robots into the manufacturing system. The system can be used to verify and correct control algorithms including robot movements and to evaluate the effectiveness of a robot and other machines in the planning stage.

A multi-computer control architecture composed a system computer and several control computers has been adopted as shown in Fig. 17. The computer control architecture was

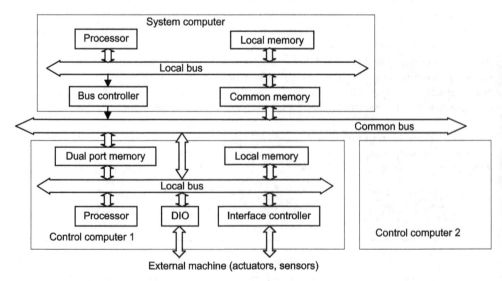

Fig. 17. Multi-computer control architecture composed a system computer and several control computers with dual port memory

developed for the use of distributed autonomous control of independent actuators or machines in compact factory automation systems (Yasuda & Tachibana, 1987). The system controller controls communication between the system controller and control computers through the bus controller based on the master-slave mode. The system computer installs the conceptual net model for system coordination and installs local net models in the control controllers through the common bus. The control computers are equipped with interface circuits to actuators and external sensors for direct machine control and monitoring. Then, in the real-time control, the system computer communicates with each control computer through dual port memory with respect to firing of shared transitions and gate arc signals (Yasuda, 2011). The presented control flow of the net model is successfully executed using output signal arc and permissive gate arc. The net model in the system controller is conceptual for system coordination and not so large. The computation speed of 50 MHz of the general microprocessor is satisfactorily high in comparison with those of controlled devices such as robotic arms, conveyors, machine tools and external sensors.

6. Conclusions

A Petri net based specification and real-time control method for large complex robotic manufacturing systems was introduced as an effective prototyping tool to realize distributed autonomous control systems corresponding to the hardware structure of robotic manufacturing systems. From the design point of view, the use of nets has many advantages in modeling, qualitative analysis, performance evaluation and code generation. The Petri net appears as a key formalism to describe, analyze and implement the distributed autonomous control system for manufacturing systems in future.

7. References

Gentina, J. C. & Corbeel, D. (1987). Coloured Adaptive Structured Petri-Net: A Tool for the Automatic Synthesis of Hierarchical Control of Flexible Manufacturing Systems, *Proceedings of 1987 IEEE International Conference on Robotics and Automation*, pp. 1166-1172

Holt, J. D. & Rodd, M. G. (1994). An Architecture for Real-Time Distributed AI-Based Control Systems. In: *IFAC Distributed Computer Control Systems 1994*, 47-52

Holding, D. J., & Sagoo, J. S. (1992). A Formal Approach to the Software Control of High-Speed Machinery. In: *Transputers in Real-Time Control*, G. W. Irwin & P. J. Fleming (Eds.), 239-282, Research Studies Press, Taunton, Somerset, U.K.

Maletz, M. C. (1983). An Introduction to Multi-robot Control Using Production Systems, *Proceedings of IEEE Workshop on Languages for Automation*, pp. 22-27

Murata, T. (1989). Petri Nets: Properties, Analysis and Applications. *Proceedings of the IEEE*, Vol. 77, No. 4, pp. 541-580

Silva, M. (1990). Petri Nets and Flexible Manufacturing. In: *Advances in Petri Nets* G. Rozenberg (Ed.), LNCS 424, 374-417, Springer-Verlag, Berlin, Germany

Taubner, D. (1988). On the Implementation of Petri Nets. In: *Advances in Petri Nets* G. Rozenberg (Ed.), LNCS 340, 419-439, Springer-Verlag, Berlin, Germany

Wang, F. & Saridis, G. N. (1990). A Coordination Theory for Intelligent Machines, *Proceedings of the 11th IFAC World Congress*, pp. 235-240

Yasuda, G. (2000). A Multiagent Control Architecture for Multiple, Cooperating Robot Systems, *Proceedings of the International Conference on Production Research Special ICPR-2000*, Paper ID 224, August, 2000

Yasuda, G. (2008). Implementation of Distributed Control Architecture for Industrial Robot Systems Using Petri Nets, *Proceedings of the 39th International Symposium of Robotics*, pp.533-538

Yasuda, G. (2010). Petri Net Based Implementation of Hierarchical and Distributed Control for Discrete Event Robotic Manufacturing Systems, *Proceedings of the 2010 IEEE International Conference on Control Applications, Part of 2010 IEEE Multi-Conference on Systems and Control*, pp.251-256

Yasuda, G. (2011). Design and Implementation of Distributed Control Architecture for Flexible Manufacturing Cells Based on Petri nets, *Proceedings of 12th Asia-Pacific Industrial Engineering & Management Systems Conference*, pp. 852-864

Yasuda, G. & Tachibana, K. (1987). A Multimicroprocessor-Based Distributed Processing System for Advanced Robot Control, *Proceedings of the IXth International Conference on Production Research*, pp.1926-1933

Permissions

The contributors of this book come from diverse backgrounds, making this book a truly international effort. This book will bring forth new frontiers with its revolutionizing research information and detailed analysis of the nascent developments around the world.

We would like to thank Engr. Dr. Faieza Abdul Aziz, Dr. Azmah Hanim Mohamed Ariff, Dr. Azfanizam Ahmad, Ir.N.Jayaseelan and Dr. Qaisar Ahsan , for lending their expertise to make the book truly unique. They have played a crucial role in the development of this book. Without their invaluable contribution this book wouldn't have been possible. They have made vital efforts to compile up to date information on the varied aspects of this subject to make this book a valuable addition to the collection of many professionals and students.

This book was conceptualized with the vision of imparting up-to-date information and advanced data in this field. To ensure the same, a matchless editorial board was set up. Every individual on the board went through rigorous rounds of assessment to prove their worth. After which they invested a large part of their time researching and compiling the most relevant data for our readers. Conferences and sessions were held from time to time between the editorial board and the contributing authors to present the data in the most comprehensible form. The editorial team has worked tirelessly to provide valuable and valid information to help people across the globe.

Every chapter published in this book has been scrutinized by our experts. Their significance has been extensively debated. The topics covered herein carry significant findings which will fuel the growth of the discipline. They may even be implemented as practical applications or may be referred to as a beginning point for another development. Chapters in this book were first published by InTech; hereby published with permission under the Creative Commons Attribution License or equivalent.

The editorial board has been involved in producing this book since its inception. They have spent rigorous hours researching and exploring the diverse topics which have resulted in the successful publishing of this book. They have passed on their knowledge of decades through this book. To expedite this challenging task, the publisher supported the team at every step. A small team of assistant editors was also appointed to further simplify the editing procedure and attain best results for the readers.

Our editorial team has been hand-picked from every corner of the world. Their multi-ethnicity adds dynamic inputs to the discussions which result in innovative outcomes. These outcomes are then further discussed with the researchers and contributors who give their valuable feedback and opinion regarding the same. The feedback is then collaborated with the researches and they are edited in a comprehensive manner to aid the understanding of the subject.

Apart from the editorial board, the designing team has also invested a significant amount of their time in understanding the subject and creating the most relevant covers. They scrutinized every image to scout for the most suitable representation of the subject and create an appropriate cover for the book.

The publishing team has been involved in this book since its early stages. They were actively engaged in every process, be it collecting the data, connecting with the contributors or procuring relevant information. The team has been an ardent support to the editorial, designing and production team. Their endless efforts to recruit the best for this project, has resulted in the accomplishment of this book. They are a veteran in the field of academics and their pool of knowledge is as vast as their experience in printing. Their expertise and guidance has proved useful at every step. Their uncompromising quality standards have made this book an exceptional effort. Their encouragement from time to time has been an inspiration for everyone.

The publisher and the editorial board hope that this book will prove to be a valuable piece of knowledge for researchers, students, practitioners and scholars across the globe.

List of Contributors

Hirohisa Narita
School of Health Sciences, Fujita Health University, Japan

Moacyr Carlos Possan Junior
Whirlpool Latin America, Brazil
Santa Catarina State University – UDESC, Brazil

André Bittencourt Leal
Santa Catarina State University – UDESC, Brazil

Hasse Nylund and Paul H Andersson
Tampere University of Technology, Finland

Alberto Regattieri
DIEM - Department of Industrial and Mechanical Plants, University of Bologna, Italy

Raad Yahya Qassim
Department of Ocean and Naval Engineering, COPPE, Federal University of Rio de Janeiro, Brazil

Michael A. Saliba and Anthony Caruana
Department of Industrial and Manufacturing Engineering, University of Malta, Malta

Gökhan Eğilmez and Gürsel A. Süer
Ohio University, USA

Orhan Özgüner
Johns Hopkins University, USA

Jing Huang and Gürsel A. Süer
Ohio University, USA

Xiaohong Suo
China Institute of Industrial Relations, Beijing, P.R.China

Gen'ichi Yasuda
Nagasaki Institute of Applied Science, Japan

Printed in the USA
CPSIA information can be obtained
at www.ICGtesting.com
JSHW011406221024
72173JS00003B/432